Stephan Rhode

Robust and Regularized Algorithms for Vehicle Tractive Force Prediction and Mass Estimation

Karlsruher Schriftenreihe Fahrzeugsystemtechnik
Band 62

Herausgeber
FAST Institut für Fahrzeugsystemtechnik
Prof. Dr. rer. nat. Frank Gauterin
Prof. Dr.-Ing. Marcus Geimer
Prof. Dr.-Ing. Peter Gratzfeld
Prof. Dr.-Ing. Frank Henning

Das Institut für Fahrzeugsystemtechnik besteht aus den Teilinstituten Bahnsystemtechnik, Fahrzeugtechnik, Leichtbautechnologie und Mobile Arbeitsmaschinen.

Eine Übersicht aller bisher in dieser Schriftenreihe erschienenen Bände finden Sie am Ende des Buchs.

Robust and Regularized Algorithms for Vehicle Tractive Force Prediction and Mass Estimation

by
Stephan Rhode

KIT Scientific Publishing

Dissertation, Karlsruher Institut für Technologie
KIT-Fakultät für Maschinenbau

Tag der mündlichen Prüfung: 29. August 2016
Referenten: Prof. Dr. rer. nat. Frank Gauterin
Prof. Dr. ir. Ivan Markovsky

Impressum

KIT Scientific Publishing

Karlsruher Institut für Technologie (KIT)
KIT Scientific Publishing
Straße am Forum 2
D-76131 Karlsruhe

www.ksp.kit.edu

Print on Demand 2018 – Gedruckt auf FSC-zertifiziertem Papier

ISSN 1869-6058
ISBN 978-3-7315-0807-6
DOI 10.5445/KSP/1000083492

Vorwort des Herausgebers

Die Fahrzeugtechnik ist gegenwärtig großen Veränderungen unterworfen. Klimawandel, die Verknappung einiger für Fahrzeugbau und -betrieb benötigter Rohstoffe, globaler Wettbewerb, gesellschaftlicher Wandel und das rapide Wachstum großer Städte erfordern neue Mobilitätslösungen, die vielfach eine Neudefinition des Fahrzeugs erforderlich machen. Die Forderungen nach Steigerung der Energieeffizienz, Emissionsreduktion, erhöhter Fahr- und Arbeitssicherheit, Benutzerfreundlichkeit und angemessenen Kosten finden ihre Antworten nicht aus der singulären Verbesserung einzelner technischer Elemente, sondern benötigen Systemverständnis und eine domänenübergreifende Optimierung der Lösungen.
Hierzu will die Karlsruher Schriftenreihe für Fahrzeugsystemtechnik einen Beitrag leisten. Für die Fahrzeuggattungen Pkw, Nfz, Mobile Arbeitsmaschinen und Bahnfahrzeuge werden Forschungsarbeiten vorgestellt, die Fahrzeugsystemtechnik auf vier Ebenen beleuchten: das Fahrzeug als komplexes mechatronisches System, die Fahrer-Fahrzeug-Interaktion, das Fahrzeug in Verkehr und Infrastruktur sowie das Fahrzeug in Gesellschaft und Umwelt.
Automatisierung und Vernetzung von Fahrzeugen eröffnen ein großes Potenzial, den Betrieb von Kraftfahrzeugen sicherer und komfortabler zu gestalten und gleichzeitig ihren Energiebedarf und die benötigte Fahrzeit zu verringern. Durch die Verwendung vorausschauender Informationen, die aus Kartenmaterial, Vernetzung mit vorausfahrenden Fahrzeugen und Backend-Servern sowie mit der eigenen Fahrzeugsensorik gewonnen werden können, kann der Betriebsablauf des Fahrzeugs online optimiert

werden. Hierfür müssen die Fahrgeschwindigkeit und die Gangstufe sowie weitere Parameter wie z.B. der Ladezustand einer Traktionsbatterie über der Fahrstrecke berechnet werden, was u.a. die genaue Vorhersage der Traktionskraft an den Rädern voraussetzt. Dafür werden geeignete Fahrzeugmodelle benötigt, die die erforderliche Genauigkeit liefern und mit der im Fahrzeug verfügbaren geringen Rechenleistung und einem kleinen Speicherplatzbedarf auskommen. Eine Herausforderung liegt dabei darin, dass nicht alle Modellparameter bekannt sind, sie sich wie die Fahrzeugmasse, der Rollwiderstands- oder der Luftwiderstandsbeiwert über der Zeit ändern können und daher während der Fahrt online geschätzt werden müssen.
Hier setzt die Arbeit von Herrn Rhode an, der Alternativen zu den üblicherweise aus Ausrollversuchen ermittelten Fahrwiderstandsmodellen mit zeitinvarianten Koeffizienten aufzeigt. Dazu beleuchtet er systematisch Möglichkeiten der Modellbildung und der Online-Parameterschätzung und erarbeitet effiziente Methoden, die trotz schlecht gestellter Probleme und der im realen Fahrbetrieb stets auftretenden Ausreißer und anregungsarmen Zeitabschnitte genaue und schnell konvergierende Ergebnisse liefern.

Karlsruhe, 29.8.2016 *Prof. Dr. rer. nat. Frank Gauterin*

Abstract

This dissertation provides novel robust and regularized algorithms from linear system identification for parameter estimation with applications in vehicle tractive force prediction and mass estimation.
Energy efficient look-ahead vehicle controllers and range prediction of electric vehicles require accurate prediction of the vehicle tractive force. Yet, precise vehicle mass estimates are fundamental in active safety assistance systems.
The combination of two linear gray-box models ($\mathcal{M}_3$ and $\mathcal{M}_4$) with unknown vehicle parameters and several (well known and novel) recursive estimators gave a set of candidate models.
Given a large record of real world data from test runs on public roads, recursive algorithms adjusted the unknown vehicle parameters under a broad variation of statistical assumptions. Additionally, the set of candidate models comprised a white-box model with a_V, b_V, and c_V parameters (abc) that represented the state of art in vehicle tractive force prediction.
The best model estimator combination in terms of vehicle tractive force prediction quality was $\mathcal{M}_4$ with the novel recursive regularized M-estimator (RRLM), depicted by cross validation. Moreover, $\mathcal{M}_4$-RRLM was significantly superior compared with the conventional abc white-box model. The best model estimator combination for vehicle mass estimation was $\mathcal{M}_4$ with the novel Stenlund-Gustafsson IV M-Kalman filter, which is an estimator for the random-walk errors-in-variables model that has not been considered in related literature for vehicle mass estimation.

Index Terms—system identification, Kalman filter, errors-in-variables (EIV), total least squares (TLS), recursive estimation, robust estimation, outliers, M-estimator, regularization, wind-up, vehicle dynamics, vehicle mass, rolling resistance, cornering resistance.

Kurzfassung

Diese Arbeit bietet neuartige robuste und regularisierte Algorithmen aus der linearen Systemidentifikation für die Parameterschätzung mit Anwendungen in der Fahrzeugradzugkraftprädiktion und Masseschätzung.
Energieeffiziente vorausschauende Betriebsstrategieregler und Reichweitenvorhersagemanager von Elektrofahrzeugen erfordern eine präzise Prädiktion der Fahrzeugradzugkraft. Eine präzise Schätzung der Fahrzeugmasse ist hingegen essentiell für aktive Sicherheitssysteme.
Die Kombination zweier linearer gray-box Modelle ($\mathcal{M}_3$ und $\mathcal{M}_4$) mit unbekannten Fahrzeugparametern und mehreren (bekannten und neuartigen) rekursiven Schätzern ergab eine Klasse von möglichen Modellen.
Die unbekannten Fahrzeugparameter wurden durch die rekursiven Schätzer auf Grundlage eines großen Datensatzes von Realmessungen von Testfahrten auf öffentlichen Straßen unter großer Variation der getroffenen statistischen Annahmen justiert. Zusätzlich wurde zu der Klasse von möglichen Modellen ein white-box Modell mit a_V, b_V , und c_V Parametern (abc) hinzugefügt, welches den bisherigen Stand in der Fahrzeugradzugkraftprädiktion darstellte.
Die Kombination aus $\mathcal{M}_4$ und dem neuartigen rekursiven regularisierten M-Schätzer (RRLM) lieferte die Modellkombination mit der genauesten Fahrzeugradzugkraftprädiktion, welche durch Kreuzvalidierung bestimmt wurde. Darüber hinaus war die Kombination aus $\mathcal{M}_4$-RRLM dem konventionellen abc white-box Modell signifikant überlegen. Die beste Modell Schätzerkombination zur Fahrzeugmassenschätzung lieferte $\mathcal{M}_4$ zusammen mit dem neuartigen Stenlund-Gustafsson IV M-Kalman Filter, welches einen

Schätzer des random-walk errors-in-variables Modells darstellt, dass bisher nicht in der Literatur zur Fahrzeugmassenschätzung berücksichtigt wurde.

Acknowledgments

I would like to gratefully and sincerely thank my adviser Frank Gauterin for his guidance, encouragement, and confidence in my work during my graduate studies at the Institute of Vehicle System Technology at Karlsruhe Institute of Technology. I am sure that I will held the record for being abroad from the institute for a long time, which would have been impossible without the support of Frank Gauterin.

In this context, I thank Karlsruhe House of Young Scientists for the various scholarships that I received for studies abroad. The PhD support through Karlsruhe House of Young Scientists was awesome and is a figurehead of PhD education. In particular, Gaby Weick and Katharina Strobel provided much support and hints so that I was able to gain these scholarships.

I am grateful for the comments of my co-adviser Ivan Markovsky, who largely influenced my research and working methods. Moreover, the collaboration with Ivan and Konstantin Usevich during my stay at the department ELEC in Vrije Universiteit Brussel was exceptionally fruitful and up to now, the most professional teamwork that I have ever experienced.

I like to thank Michael Frey for his support and inspiring ideas for this dissertation. Not to forget his effort in keeping me away from university bureaucracy, which is a challenging example for a time-varying black-box system.

This work was accomplished in cooperation with the Energy Management Complete Vehicle Department of Porsche and I am grateful for the experience within the project *Energy Efficient Driving Strategy.* Thanks to Frank Weberbauer for his guidance, the interesting discussions, and for providing

test vehicles. I have to say that I really enjoyed the test-runs and some of them were probably redundant ;)
Special thanks to Philip Markschläger. Philip was my main contact person at Porsche and supported me with data, information, open discussions, and sometimes funny stories from the car industry. I enjoyed to work with Philip.
I would also like to thank Markus Goslar and Johannes Bach of Forschungszentrum Informatik. I still remember my desperate trials to overcome some Simulink quirks at the beginning and the exceptional patience of Master Markus during my "hello world" Padawan lessons. Johannes spent countless hours in developing a vehicle signal recorder and converter according to my special needs. The work of Johannes was vital to handle the enormous amount of real world data within this dissertation.
Many student researchers supported this dissertation and I want to emphasize the outstanding work and support of Markus Knobloch, Harry Hamann and Florian Mutter. Markus and Florian were exceptional reliable students and I am not sure who learned more during my supervision, the students or the supervisor? The master's thesis and conference paper of Harry established the contact to Karl Hedrick and many researchers of the vehicle dynamics lab at University of California, Berkeley, where I spent a semester abroad and worked with Sanghyun Hong on a second collaborative publication.
I will probably fail to acknowledge the support of Felix Bleimund appropriately within a few lines, but I will give my best. Felix was like me a geek[1] in our institute and influenced my research significantly. I am sure that neither his nor mine dissertation would contain the grade of novelty without knowing each other. I remember the period when we investigated the polynomial Kalman smoother, which was the first time that I felt a flow during collaborative work.

[1] Let me define geek loosely as expert or enthusiast that is obsessed by a technical challenge. Do not mix up geek with nerd, which is a social impaired geek.

I am thankful for the comments of Moritz Vaillant, who spent many hours to read the first draft of this dissertation. Moreover, special thanks to Larissa Fritzenschaf for her patient help to fix many typos and punctuation errors. Finally, my parents receive my deepest gratitude for their dedication and the many years of support during my studies that provided the foundation for this work.
I am sure that I have forgotten someone and apologize for that.

Berkeley, October 2014 *Stephan Rhode*

Acronyms

abc	white-box model with a_V, b_V, and c_V parameters
ARX	auto regressive with eXogenous input
BP	backpropagation
CAN	control area network
CG_v	vehicle center of gravity
CG_W	wheel center of gravity
EIV	errors-in-variables
EKF	extended Kalman filter
FIR	finite impulse response
GTLS	generalized TLS
IAE	innovation-based adaptive estimation
IC	instantaneous center
IMME	interacting multiple model estimation
IR	impulse response
IV	instrumental variables
IVKF	IV Kalman filter
IVMKF	IV M-Kalman filter
KF	Kalman filter
LMS	least median of squares

LS	least squares
LTS	least trimmed squares
MISO	multi-input-single-output
MKF	M-Kalman filter
MME	multiple model estimation
NCE	noise covariance estimator
NN	neural networks
OE	output-error
PF	polynomial-function
PKS	polynomial Kalman smoother
PP	pressure point
PSD	predictive route data
RGTLS	recursive GTLS
RIV	recursive IV
RIVM	recursive IV M-estimator
RIVMKF	regularized IV M-Kalman filter
RLM	recursive M-estimator
RLS	recursive least squares
RLSmf	recursive least squares with multiple forgetting
RMKF	regularized M-Kalman filter
RRIVM	recursive regularized IV M-estimator
RRLM	recursive regularized M-estimator
RTIV	recursive total instrumental variables
RTLS	recursive TLS
RW	random-walk
SGF	Savitzky Golay filter
SGIVMKF	Stenlund-Gustafsson IV M-Kalman filter

SGMKF	Stenlund-Gustafsson M-Kalman filter
SISO	single-input-single-output
STSP	state-space
TLS	total least squares
UKF	unscented Kalman filter
WLS	weighted least squares

Symbols

Vehicle dynamics

A_{B} (m^2)	brake piston cross-sectional area
α_{W} (rad)	slip angle
A_{V} (m^2)	vehicle cross-sectional area
a_{V} (N)	vehicle a parameter
β_{V} (rad)	vehicle side slip angle
b_{V} ($\mathrm{kg\,s^{-1}}$)	vehicle b parameter
c_{V} ($\mathrm{kg\,m^{-1}}$)	vehicle c parameter
c_w	$c_x(\psi_{\mathrm{a}} = 0)$
c_x	longitudinal drag coefficient
$c_{x\mathrm{W}}$ (N)	wheel longitudinal stiffness
c_y	lateral drag coefficient
$c_{y\mathrm{W}}$ ($\mathrm{N\,rad^{-1}}$)	wheel cornering stiffness
δ_{S} (rad)	steering wheel angle
δ_{W} (rad)	wheel steer angle
e_{R} (m)	distance between $F_{z\mathrm{W}}$ und z_{W}
F (N)	force
f_{r}	coefficient of rolling resistance
F_{ac} (N)	acceleration force
F_{cl} (N)	climbing force
F_{c} (N)	centrifugal force
F_{W} (N)	wheel force
F_{Wc} (N)	cornering resistance

F_{Wr} (N)	rolling resistance
F_{Wt} (N)	toe-in resistance
$F_{x_{\mathrm{a}}}$ (N)	longitudinal aerodynamic resistance
$F_{x\mathrm{V}}$ (N)	vehicle tractive force
$F_{x\mathrm{W}}$ (N)	wheel longitudinal force
$F_{y_{\mathrm{a}}}$ (N)	lateral aerodynamic resistance
$F_{y\mathrm{W}}$ (N)	wheel lateral force
$F_{z\mathrm{W}}$ (N)	wheel vertical force
G	gear
g (m s^{-2})	gravitational constant
I_{B} (kg m^2)	braking moment of inertia
I_{C} (kg m^2)	clutch moment of inertia
I_{D} (kg m^2)	differential moment of inertia
i_{D}	differential ratio
I_{E} (kg m^2)	engine moment of inertia
I_{G} (kg m^2)	gearbox moment of inertia
i_{G}	gearbox ratio
I_{red} (kg m^2)	reduced moment of inertia
I_{W} (kg m^2)	wheel moment of inertia
$I_{z\mathrm{V}}$ (kg m^2)	vehicle yaw moment of inertia
j_{W}	wheel index: 1-rear left, 2-rear right, 3-front right, …
l (m)	wheel base
μ_{W}	wheel friction coefficient
$\mu_{x\mathrm{W}}$	longitudinal wheel friction coefficient
$\mu_{y\mathrm{W}}$	lateral wheel friction coefficient
m_{V} (kg)	vehicle mass
p_{a} (Pa)	air pressure
p_{B} (Pa)	braking pressure
ϕ_{r} (rad)	road bank angle
ϕ_{V} (rad)	vehicle roll angle
ϕ_{W} (rad)	wheel camber angle

ψ_{a} (rad)	air approach angle
ψ_{V} (rad)	vehicle yaw angle
ψ_{w} (rad)	wind angle
r (m)	path radius
R_{a} (J kg^{-1} K^{-1})	specific gas constant
r_{B} (m)	braking radius
ρ_{a} (kg m^{-3})	air density
r_{r} (m)	road curvature radius
r_{W} (m)	dynamic wheel radius
s_{W}	wheel slip
$s_{x\mathrm{W}}$	longitudinal wheel slip
$s_{y\mathrm{W}}$	lateral wheel slip
θ (rad)	gradient angle
θ_{D} (rad)	differential shaft rotation angle
θ_{E} (rad)	engine rotation angle
θ_{G} (rad)	gear shaft rotation angle
θ_{r} (rad)	road angle
θ_{V} (rad)	vehicle pitch angle
θ_{W} (rad)	wheel rotation angle
T_{a} (K)	air temperature
T_{B} (N m)	braking torque
T_{D} (N m)	differential input torque
T_{E} (N m)	engine torque
T_{G} (N m)	gearbox input torque
T_{R} (N m)	rim torque
v_{a} (m s^{-1})	air approach velocity
v_{V} (m s^{-1})	vehicle velocity, $v_{\mathrm{V}} = \sqrt{\dot{x}_{\mathrm{V}}^2 + \dot{y}_{\mathrm{V}}^2 + \dot{z}_{\mathrm{V}}^2}$
v_{W} (m s^{-1})	wheel velocity, $v_{\mathrm{W}} = \sqrt{\dot{x}_{\mathrm{W}}^2 + \dot{y}_{\mathrm{W}}^2 + \dot{z}_{\mathrm{W}}^2}$
v_{w} (m s^{-1})	wind velocity
x_0 (m)	gravity-fixed longitudinal axis
X_{B}	lumped braking parameter

x_{r} (m)	road longitudinal axis
x_{V} (m)	vehicle longitudinal axis
X_{W}	cornering stiffness
x_{W} (m)	wheel longitudinal axis
y_0 (m)	gravity-fixed lateral axis
y_{r} (m)	road lateral axis
y_{V} (m)	vehicle lateral axis
y_{W} (m)	wheel lateral axis
z_0 (m)	gravity-fixed vertical axis
z_{r} (m)	road vertical axis
z_{V} (m)	vehicle vertical axis
z_{W} (m)	wheel vertical axis

System identification

A	measured input
$\mathcal{A}$	state matrix
$\widehat{A}$	estimated input
AIC	Akaike's information criterion
$\mathfrak{A}$	instruments
$\widetilde{A}$	input noise
$\breve{A}$	estimated input noise
$\overline{A}$	true input
B	measured output
$\mathcal{B}$	input matrix
β	Myriad tuning constant
$\widehat{B}$	estimated output
BIC	Bayesian information criterion
$\widetilde{B}$	output noise
$\breve{B}$	estimated output noise
$\overline{B}$	true output

C	Cholesky factor of $\widetilde{P}$
$\boldsymbol{C}$	output matrix
c	condition number
$\mathcal{D}$	feed-through matrix
d	number of outputs
δ	Huber tuning constant
ΔA	input correction
ΔAIC	AIC difference
ΔB	output correction
ΔX	parameter correction
ΔZ	augmented correction
η	learn rate
I	identity matrix
k	number of prediction steps
κ	regularization parameter
$\mathcal{L}$	cost function
L	correction vector
λ	forgetting factor
$\mathcal{M}$	model
m	samples
MAD	median absolute deviation
MEDSE	median squared error
MSE	mean squared error
n	number of inputs
NMSE	normalized mean squared error
NRMSE	normalized root mean squared error
o	number of estimable parameters
P	covariance matrix
p	probability
P_{d}	desired P
$\widetilde{P}$	noise covariance matrix

$\breve{P}$	estimated noise covariance matrix
ψ	influence function
$\mathcal{Q}$	covariance of parameter correction
q	$n + d$
$\widehat{\mathcal{Q}}$	estimated quantile
$\mathcal{Q}$	quantile
R	input covariance matrix
$\mathcal{R}$	regularization matrix
R^1	covariance of output noise
ρ	ρ-function
S	matrix of eigenvalues
SEVN	squared error vector norm
σ	scale
$\widehat{\sigma}$	estimated scale
t (s)	time
τ (s)	delay
U	matrix of left-singular vectors
V	matrix of eigenvectors
W	scaling matrix
w_l	left window
w_r	right window
W_l	left scaling matrix
W_r	right scaling matrix
W_X	parameter scaling matrix
X	parameter
$\widehat{X}$	estimated parameter
$\widehat{X_\mathrm{c}}$	constrained estimated parameter
X_max	upper bound
X_min	lower bound
$\overline{X}$	true parameter
Z	augmented data

$\widehat{Z}$	estimated augmented data
$\check{Z}$	augmented estimated noise

Mathematical operators

$\mathrm{chol}(\cdot)$	Cholesky factorization
$\mathrm{cov}(\cdot)$	covariance operator
$\mathrm{diag}(\cdot)$	diagonal elements
$\mathbb{E}$	expectation operator
Λ	shift operator
$\mu(\cdot)$	arithmetic mean
$\mathrm{med}(\cdot)$	median
$\mathcal{N}$	Gaussian distribution
$\lVert\cdot\rVert_1$	1-norm
$\lVert\cdot\rVert_2$	Euclidean norm
$\lVert\cdot\rVert_F$	Frobenius norm
$\odot$	element-wise product
$\mathbb{R}$	rational numbers
sgn	signum
$\mathrm{svd}(\cdot)$	singular value decomposition
$\mathbb{Z}$	integer numbers

Contents

1 Introduction

"publish or perish"

Wilson [192, p. 197] [see 60].

Outline

This chapter gives the aim, motivation, and research questions of this dissertation and introduces energy efficient vehicle look-ahead controllers as primary application. One section gives the outline of this dissertation and hints how to read it most efficiently. The following two sections discuss methods that ensure reproducible research and introduce specific typography to avoid self-plagiarism.

1.1 Aim of this work and research questions

The main objective of this work is to induce models that give accurate prediction of the vehicle tractive force. The vehicle tractive force is the force that is required to operate a vehicle on desired speed. This force acts in vehicle longitudinal direction and may be positive or negative. A positive vehicle tractive force means that the engine propels the vehicle. Hence, the engine torque is positive and the engine consumes fuel or state of charge. A negative vehicle tractive force requires to slow down the vehicle with brakes or an electric engine that operates as generator. A precise definition of the vehicle tractive force is given in Definition 1.1.

Definition 1.1. The vehicle tractive force is the force that balances the force equilibrium of internal resistances and external forces and resistances.

The main application of the introduced vehicle tractive force models is in look-ahead vehicle controllers that reduce the propel energy of vehicles [131, pp. 11–18], [12, pp. 47–49] and [59, 75, 90, 188, 189]. Basically all look-ahead vehicle controllers require the precise acquisition and prediction of the vehicle tractive force for a forthcoming horizon of the route.

Generally speaking, a look-ahead controller is a control scheme that uses predictive (future) information and an optimization method to compute the control action for the plant. One branch of look-ahead controllers is model predictive control that is also known as receding horizon control and very successful in automotive and many other industry applications [130]. Another branch is dynamic programming [17].

Another application of the introduced vehicle tractive force models is range prediction for electric vehicles [43, 66] that relies on highly accurate vehicle tractive force prediction. Moreover, the emerging field of autonomous driving [197, Figure 12] requires inverse vehicle dynamics models for longitudinal and lateral controllers.

Only information from existing standard built-in vehicle sensors will be used, in order to determine a cost-effective and quickly applicable method. Accordingly, the presented methods are applicable to most standard vehicles without further ado.

The first research question is thus: *how can we produce, while driving, highly precise vehicle tractive force models with little computational effort and without additional sensors?*

We will see in Section 2.8 that we can derive vehicle tractive force models with different model structure (different model complexity) and we can choose from different estimators which follow different statistical assumptions.

Hence, we can ask a second research question: *Which combination of model structure and estimator gives the most accurate prediction of the vehicle tractive force?*

Given the variety of possible model structures and estimators a third research question and hence, a second objective of this work arises: *which model estimator combination yields superior accurate vehicle mass estimates?* This question is important because the vehicle mass is one of the most significant parameters in active safety systems such as the anti-lock braking system or electronic stability program.

1.2 Motivation

Vehicle longitudinal dynamics are based on an equilibrium between vehicle propulsion, internal resistances, and external resistances and forces. The internal resistances are rather independent of surrounding conditions, compared with the external resistances and forces. A sufficiently accurate characterization of the internal resistances is given by results of test bed measurements that give data to develop drive-train models. Hence, the focus of this work will be the modeling of external resistances and forces by use of model-based estimators.

Chapter 2 will group the various parts of the external resistances and forces as climbing force, acceleration force, longitudinal aerodynamic resistance, rolling resistance, cornering resistance, and lateral aerodynamic resistance. The external resistances and forces depend on lumped parameters which vary more or less quickly while driving and depend strongly on the current environmental conditions.

The equilibrium between vehicle propulsion, internal resistances and external resistances and forces causes that a desired driving state can only be achieved if enough propulsive power is available, resistances are decreased or energy is reused to propel the vehicle. Concerning limited resources, a focus of automotive engineering is energy efficient driving, which can be achieved by improving the components and subsystems of the vehicle (reduce resistances, add components which recuperate and store energy) or by an intelligent operation strategy of the vehicle (reuse energy). The development of individual components like engine or gearbox is well advanced, whereas the use of intelligent operation strategies, such as look-ahead control, offers potential to achieve high driving performance on low propel energy.

1.3 How to read this dissertation

This outline will give an idea how to read this work and what to expect from each chapter. Suggestions are given which parts of this work may be preliminary skipped depending on the background of the reader. I highly recommend to read the pdf version of this work, because each citation, symbol and acronym has a hyperlink to its reference, definition or description, respectively. Moreover, all figures are shown as vector graphics. Hence, you can zoom each figure to study the details without loosing image quality.

Chapter 2 — Vehicle force equilibrium

contains fundamentals in vehicle dynamics. Although numerous books cover this topic in more detail, it is useful to summarize internal and external resistances and forces that act on vehicles to ensure a consistent nomenclature.
The introduced concepts are fundamental to follow Chapter 4 and Chapter 5. Readers who are familiar with vehicle dynamics may skip Chapter 2 and refer back for a definition of vehicle specific symbols if needed.
However, Section 2.8 may also be of interest for vehicle dynamics experts, because Section 2.8 gives the link between vehicle dynamics and concepts from system identification which are covered in Chapter 3.

Chapter 3 — Models and estimators

gives a broad survey of methods in linear model structures and estimators. This chapter is fundamental to follow all subsequent chapters (Chapter 4–

Chapter 6) which are rather vehicle specific. Moreover, Chapter 3 defines general concepts such as system, model, model structure, and experimental condition. Additionally, this chapter classifies numerous model structures and estimators in Figure 3.1 which is a well arranged tree diagram. Methods like model selection, model validation, cross-validation, regularization, and robustness will be introduced in detail.
The main part of this chapter is a survey of various statistical models and their estimators which will be shown in numerous batch and recursive algorithms and explored by reproducible numerical experiments.
This chapter highlights connections and transitions between well known algorithms and explores novel recursive estimators that are applied with real world data in Chapter 5. Additional topics and a hands on guideline for choosing the most appropriate estimator concludes this chapter.
This chapter may be of interest for a broad range of readers from engineering or more general for readers with interest in parameter estimation. The connection between methods from this chapter and vehicle science follows in Chapter 4 and Chapter 5. Sections that are marked with an asterisk comprise rather complex mathematical content or are of more theoretical interest. These sections may be skipped from practitioners at first reading.

Chapter 4 — Survey of related research

reviews related research of vehicle parameter estimation and discovers open topics addressed by this dissertation in Chapter 5.

Chapter 5 — Vehicle tractive force prediction and mass estimation

applies estimators of Chapter 3 on two gray-box models which are based on a force equilibrium of Chapter 2. The prediction quality for the vehicle tractive force of several recursive algorithms will be compared to the benchmark with a given white-box model on a large set of vehicle real world data from

test runs on public roads. The second topic will be vehicle mass estimation which is not the main focus of this dissertation but an important research field. The set of gray-box candidate models and estimators will be compared with respect to highly accurate estimates of the vehicle mass.
This chapter may be of interest for readers from vehicle science who deal with look-ahead controllers or estimators for vehicle parameters. Additionally, this chapter may give new insights for readers from system identification community due to the large amount of real world data and the experienced pitfalls and hazards that arise with challenging real world data.

Chapter 6 — Concluding remarks

provides conclusions, lists the main contributions, and gives an overview of open topics which might develop into interesting research projects for the future.

Chapter 6.3 — Algorithms

provides a link to access the supplementary material of this dissertation and shows how to execute the algorithms from Chapter 3 in Matlab®.

1.4 Some thoughts about reproducible research

Following the guidelines of [187] and the advice of my co-adviser Ivan Markovsky, I started to publish code as additional material of my papers. This dissertation is a mixture of computer experiments that show and evaluate various methods with artificial data (Chapter 3) and real world experiments with vehicles on public roads (Chapter 5). These experiments require Matlab® software and the implementation from the supplementary material of this dissertation (see Chapter 6.3). All required steps to rerun the experiments will be explained.

However, some parts of an engineer's research is simply not fully reproducible. Particularly, real world data from test runs on public roads. The environmental conditions of these test runs are not reproducible. Moreover, the vehicles we used were disposed in the meanwhile. Accordingly, the results of Chapter 5 are not fully reproducible.

1.5 Typography and matrix notation

Good typography ensures that the amount of information that the reader catches is as big as possible. Hence, I gave my best to use typography wisely and this section explains some specific typography of this dissertation.

The ✎ pencil-icon indicates that further research should be conducted. Commonly, this outlook for open research topics is given at the end of a dissertation but I think it is also helpful to depict open topics directly when they appear because explanation can be given deeper in main chapters than in Section 6.3 (Open problems).

The 🚶 pedestrian-icon in Chapter 3 highlights new findings, precisely novel estimators and algorithms.

Nowadays, rapid publication of research is vital for reputation and impact in the scientific community. A modern way to deal with rapid publication is a cumulative dissertation. However, a cumulative dissertation is quite inflexible to rearrange material or put in new findings. That is why I decided to follow a way in between (classic full written dissertation and cumulative dissertation) and inserted parts of reused material.

———— *Reused material:* The reused material in Chapter 3 appears like this paragraph. Thin rules show clearly where the reused material begins and ends. The last sentence is a citation with the comment [This reused material has been reformatted for uniformity. xx]. ————

Reformatted means that there a slight modifications in the symbols to ensure a consistent nomenclature.

Note that quotations are given in a common way with large margins on both sides and smaller font.

> This text is set as block-quotation and its appearance differs from own reused material.

Moreover, you will find the mentioned computational experiments within gray shaded boxes in Chapter 3.

> Experiment
> ...

Finally, the matrix indexing within this work is as follows. The first index of a matrix denotes rows, the second columns. For instance, $A_{i,j}$ denotes an entry at row i and column j of matrix A. The colon operator (:) denotes all rows or columns, or a range. The notation $A_{:,j}$ means all rows of column j. Conversely, $A_{i,:}$ denotes all columns of row i. However, mostly the short hand notation A_i is used instead of $A_{i,:}$. The colon operator in the index of $A_{1:3,j}$ extracts a column vector of row one–three at column j of matrix A.

Summary

This chapter provided information about the research topic and its application. Further, specific typography and matrix notation were introduced. Depending on the background of the reader, the presented outline should be a good guide where to start with reading. The next chapter will give fundamentals in vehicle dynamics.

2 Vehicle force equilibrium

Outline

This chapter gives fundamentals in tire-road contact, external resistances and forces that act on the vehicle, and the lossy torque transmission from engine to rim in terms of a drive-train model. The last section converts force equilibrium equations into a matrix form which is appropriate for several model structures. This conversion is required to follow Chapter 4 and Chapter 5.

2.1 Coordinate systems

We need several coordinate systems to locate the vehicle in space. Conversely to [114, p. 3] and [1], the inertial system x_0, y_0, z_0 in Figure 2.1 is orientated in a way that the area spanned between x_0 and y_0 is perpendicular to the vector of gravity (g). Through the large earth radius, the orientation of the inertial system changes only slightly during drive. Therefore, we can assume a quasi stationary inertial system.
The road coordinate system is defined by x_r, y_r and z_r. The inclination of the road surface with respect to g is expressed by road bank angle (ϕ_r) and the road angle (θ_r). A description of the rotational movement around z_r is not essential and therefore omitted.
Four wheels are rolling on the road surface. The wheel coordinate systems have their origin in the respective wheel center of gravity ($\mathrm{CG_W}$), assuming ideal symmetric mass distribution for each wheel and wheel vertical axis (z_W) being normal to the road surface. Beside the three translational wheel

degrees of freedom x_W, y_W and z_W, we need the wheel camber angle (ϕ_W), wheel rotation angle (θ_W), and the wheel steer angle (δ_W).
The body fixed vehicle coordinate system x_V, y_V and z_V acts in the vehicle center of gravity (CG_v). The three angles are labeled with vehicle roll angle (ϕ_V), vehicle pitch angle (θ_V), and vehicle yaw angle (ψ_V).
According to [1], CG_v moves relative to the stationary system x_0, y_0, z_0 with the vehicle velocity (v_V). Depending on the driving situation, we have to carry out certain coordinate transformations to obtain the vehicle velocity components in x_V, y_V and z_V. In the easiest case of driving straight ahead without acceleration on a horizontal road surface we can adjust $\phi_r = \theta_r = 0$, $\phi_V = \theta_V = \psi_V = 0$ and therefore the vehicle velocity components become $\dot{x_V} = v_V$ and $\dot{y_V} = 0$.

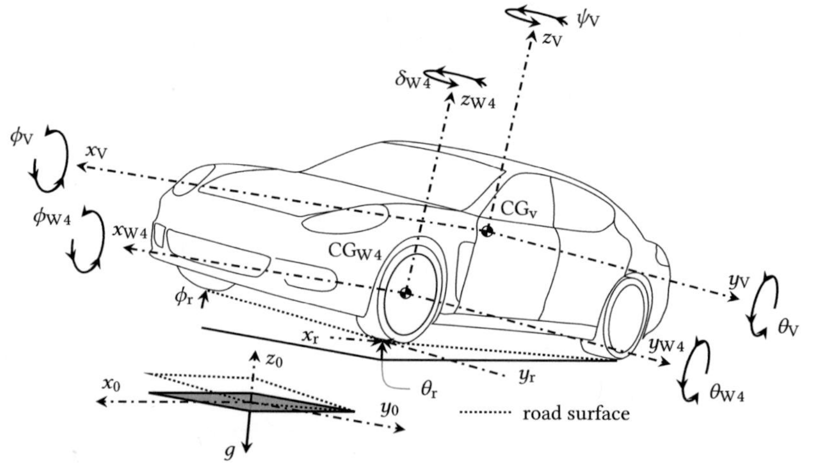

Figure 2.1: Vehicle with three translational (x_V, y_V, z_V) and three rotational degrees of freedom (ϕ_V, θ_V, ψ_V) which act in the body fixed vehicle center of gravity (CG_v). The directions x_0, y_0 of the inertial system create a surface which is perpendicular to g. The road coordinate system (x_r, y_r, z_r) is tilted with respect to this perpendicular surface around the angles ϕ_r and θ_r. The directions x_{W4}, y_{W4} and z_{W4} create one of four wheel coordinate systems which are numbered with index j_W.

2.2 Tire-road contact

Friction ensures the force transmission between tire and road surface based on the relation

$$F_W = \mu_W F_{z_W}.$$

The tire in Figure 2.2 is turned-in around δ_W and moves with the velocity v_W, whereby v_W is twisted around the slip angle (α_W). This results in two force and velocity components F_{x_W}, F_{y_W}, $\dot{x_W}$, $\dot{y_W}$, respectively, with the relations

$$\begin{aligned}
v_W &= \sqrt{\dot{x}_W^2 + \dot{y}_W^2},\\
F_W &= \sqrt{F_{x_W}^2 + F_{y_W}^2},\\
F_{x_W} &= \mu_{x_W} F_{z_W},\\
F_{y_W} &= \mu_{y_W} F_{z_W},\\
\mu_W &= \sqrt{\mu_{x_W}^2 + \mu_{y_W}^2}.
\end{aligned}$$

Note that the friction coefficients (μ_{x_W}, μ_{y_W}) depend on slip. Friction is partitioned in static friction (sticking) and kinetic friction (sliding). The deformation of the elastic tread blocks causes slip during sticking and a fraction of slip during sliding (assuming that the wheel still rotates). If static friction turns into kinetic friction, additional relative movements between the tread blocks and the road surface occur. Hence, kinetic friction causes higher slip than static friction.
Following [114, p. 31] the lateral wheel slip (s_{y_W}) is defined as

$$s_{y_W} = \tan\alpha_W = \frac{v_W \sin\alpha_W}{v_W \cos\alpha_W} = \frac{\dot{y_W}}{\dot{x_W}}.$$

However, s_{y_W} is only of minor importance in practice [114, p. 33]. Instead, α_W is directly considered to determine the wheel lateral force.

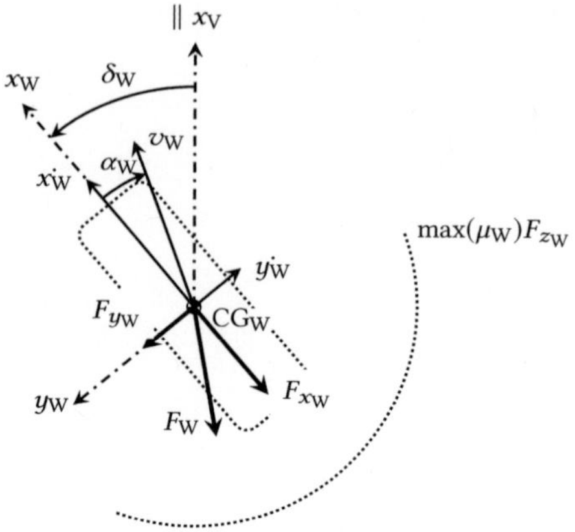

Figure 2.2: Wheel longitudinal force (F_{x_W}) and wheel lateral force (F_{y_W}) in the tire road surface contact. Kamm's circle ($\max(\mu_W)F_{z_W}$) saturates the wheel force (F_W). Accordingly, $F_W < \max(\mu_W)F_{z_W}$. Note that x_V is parallel shifted from CG_v into CG_W.

There are certain definitions for the longitudinal wheel slip (s_{x_W}) . The piecewise defined relation (2.1) ensures that s_{x_W} remains within the interval of $0 \leq s_{x_W} \leq 1$.

$$s_{x_W} = \begin{cases} \frac{\dot{\theta_W} r_W - \dot{x_W}}{\dot{\theta_W} r_W} & \dot{\theta_W} r_W > \dot{x_W} \quad \text{propellingslip} \\ \frac{\dot{x_W} - \dot{\theta_W} r_W}{\dot{x_W}} & \dot{x_W} > \dot{\theta_W} r_W \quad \text{brakeslip} \end{cases} \tag{2.1}$$

$\dot{\theta_W}$ is the wheel speed and r_W the dynamic wheel radius. The definition for s_{x_W} by [122, p. 65] and [1] is given in (2.2). The longitudinal wheel slip becomes $s_{x_W} = -1$ while braking with locked tires and can increase to infinity while propelling

$$s_{x_W} = \frac{\dot{\theta_W} r_W - \dot{x_W}}{\dot{x_W}}. \tag{2.2}$$

In the following we use solely the slip definition of (2.2).

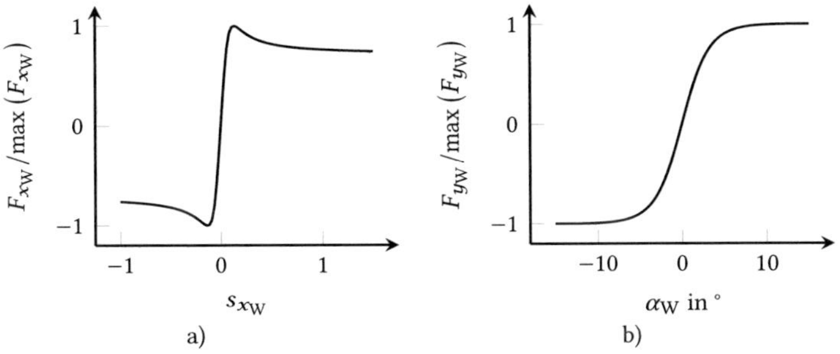

Figure 2.3: Wheel longitudinal force ($F_{x\mathrm{W}}$) as function of longitudinal wheel slip ($s_{x\mathrm{W}}$) and the wheel lateral force ($F_{y\mathrm{W}}$) as function of slip angle (α_W) generated by the magic tire formula in (2.4).

2.3 Tire models

Tire models provide mathematical functions which describe the forces and torques in the tire road surface contact. Figure 2.3 shows typical characteristics for the functions $F_{x\mathrm{W}} = f(s_{x\mathrm{W}})$ and $F_{y\mathrm{W}} = f(\alpha_\mathrm{W})$. Both curves can be approximated with a tangent around the zero point. This approach gives simple linear tire model

$$F_{x\mathrm{W}} \approx c_{x\mathrm{W}} s_{x\mathrm{W}}, \tag{2.3a}$$

$$F_{y\mathrm{W}} \approx c_{y\mathrm{W}} \alpha_\mathrm{W}, \tag{2.3b}$$

in which the two parameters $c_{x\mathrm{W}}$ and $c_{y\mathrm{W}}$ are the wheel longitudinal stiffness and wheel cornering stiffness, respectively. However model (2.3) is only valid in a narrow domain of definition around $s_{x\mathrm{W}} \approx 0$ and $\alpha_\mathrm{W} \approx 0$. The curves of $F_{x\mathrm{W}}$ and $F_{y\mathrm{W}}$ show distinct inflexion points outside of the linear domain.

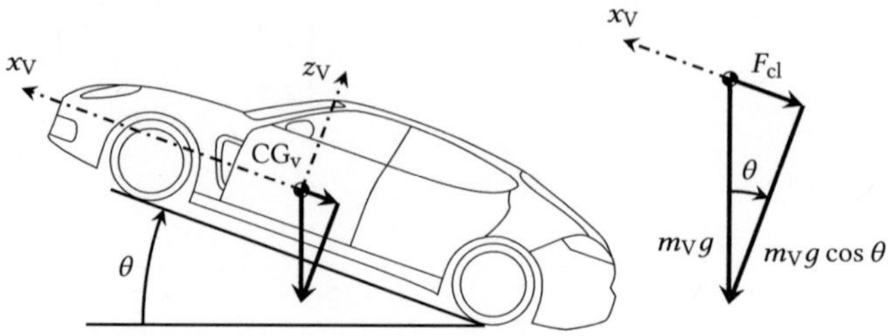

Figure 2.4: Vehicle driving uphill. The forces that act in CG_V are shown with magnification on the right.

Equation (2.4) shows a non-linear tire model which is known as magic tire formula [122, p. 165].

$$B = X_3 \sin\Big(X_2 \arctan\big(X_1 A - X_4 \left(X_1 A - \arctan\left(X_1 A\right)\right)\big)\Big) \tag{2.4}$$

The measured output (B) can be $F_{x\mathrm{W}}$ or $F_{y\mathrm{W}}$. Analogously, the measured input (A) is $s_{x\mathrm{W}}$ or α_W. The parameters ($X_1 \dots X_4$) are determined by experiments and specify the form of the function. Another often applied non-linear tire model was introduced by Burckhardt [26].

2.4 Longitudinal vehicle forces

The following forces act in vehicle longitudinal axis (x_V) and apply in vehicle center of gravity or in the pressure point (PP). The pressure point is a point where all external aerodynamic forces can be combined in a single vector.

2.4.1 Climbing force

Figure 2.4 shows a vehicle climbing a hill. The climbing force (F_cl) in (2.5), which acts in $\mathrm{CG_V}$, is determined by the vehicle mass (m_V), g and the gradient angle (θ) [69, p. 155].

$$F_\mathrm{cl} = m_\mathrm{V} g \sin\theta \tag{2.5}$$

While climbing a hill, the vehicle saves potential energy which is mainly transformed into kinetic energy (also partly transformed into electric and thermal energy) while driving downhill.

2.4.2 Acceleration force

Newton's first law states that the translational acceleration of a mass causes a acceleration force (F_{ac}) which acts contrary to the direction of acceleration.

$$F_{\mathrm{ac}} = m_{\mathrm{V}} \dot{v}_{\mathrm{V}} \tag{2.6}$$

Commonly, the relation in (2.6) comprises the reduced mass of rotational parts with $F_{\mathrm{ac}} = \left(m_{\mathrm{V}} + I_{\mathrm{red}}(G)/r_{\mathrm{W}}^2\right)\dot{v}_{\mathrm{V}}$. However, the drive-train model model in Section 2.7 will consider the rotational parts.

2.4.3 Longitudinal aerodynamic resistance

Figure 2.5 shows the longitudinal aerodynamic resistance ($F_{x_{\mathrm{a}}}$) which acts in the pressure point and originates through circulation and perfusion while driving straight ahead. In this case, $v_{\mathrm{V}} = \dot{x}_{\mathrm{V}}$. The vehicle side slip angle (β_{V}) must be considered if the vehicle drives through corners. The dimensionless longitudinal drag coefficient (c_x) is determined with wind tunnel experiments which give the function c_x over the air approach angle (ψ_{a}) ($c_x = f(\psi_{\mathrm{a}})$). Furthermore, the value on the ordinate $c_x(\psi_{\mathrm{a}} = 0)$ is often written as c_w in literature. The longitudinal aerodynamic resistance becomes

$$F_{x_{\mathrm{a}}} = \frac{\rho_{\mathrm{a}}}{2} A_{\mathrm{V}}(\psi_{\mathrm{a}}) v_{\mathrm{a}}^2 c_x(\psi_{\mathrm{a}}), \tag{2.7}$$

with the air density (ρ_{a}), vehicle cross-sectional area (A_{V}) and the air approach velocity (v_{a}) [69, p. 154]. The air density and the air approach velocity are affected by ambient conditions. If air is regarded as a dry ideal gas, the air density can be computed with the specific gas constant (R_{a}) $R_{\mathrm{a}} = 287.058\,\mathrm{J\,kg^{-1}\,K^{-1}}$ by $\rho_{\mathrm{a}} = p_{\mathrm{a}}/(R_{\mathrm{a}} T_{\mathrm{a}})$.

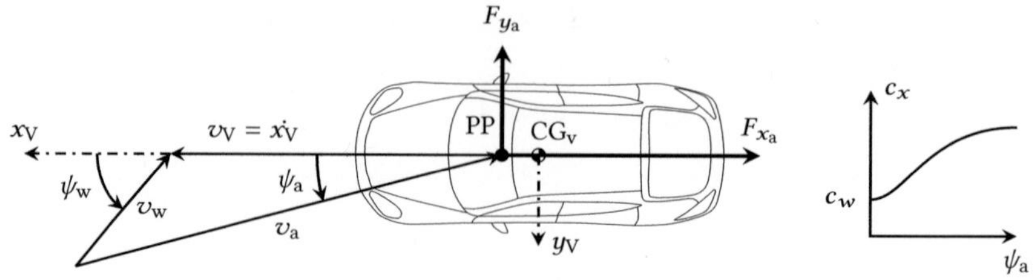

Figure 2.5: Vektor superposition for the air approach velocity (v_a) for straight ahead driving [see 69, p. 153]. The common shape of the function $c_x = f(\psi_\text{a})$ is shown on the right [see 69, p. 155]

The calculation of the exact air density is considerably more complex and exemplified in [127]. This calculation needs knowledge about the air pressure (p_a), air temperature (T_a), molar masses of water, water vapor, a compression factor, and a molar gas constant.
According to [69, p. 153], air approach velocity is given for straight driving in (2.8b) by summing up the vectors $\dot{x}_\text{V}$ and wind velocity (v_w) and considering the wind angle (ψ_w). Note that even with solely side-wind ($\psi_\text{w} = 90°$), the air approach velocity is larger than $\dot{x}_\text{V}$.

$$\vec{v_\text{a}} = \begin{pmatrix} \cos(\psi_\text{w})v_\text{w} - \dot{x}_\text{V} \\ \sin(\psi_\text{w})v_\text{w} \end{pmatrix} \tag{2.8a}$$

$$v_\text{a} = \sqrt{\dot{x}_\text{V}^2 + v_\text{w}^2 + 2\dot{x}_\text{V}v_\text{w}\cos(\psi_\text{w})} \tag{2.8b}$$

2.4.4 Rolling resistance

Haken [69, p. 150] explains that the rolling resistance (F_Wr) causes approximately 80 % of all resistances at the wheel while driving straight ahead on a dry and paved road. However, this reported fraction of F_Wr depends strongly on air approach velocity. The analogous model of the wheel in Figure 2.6 exemplifies the origin of F_Wr.

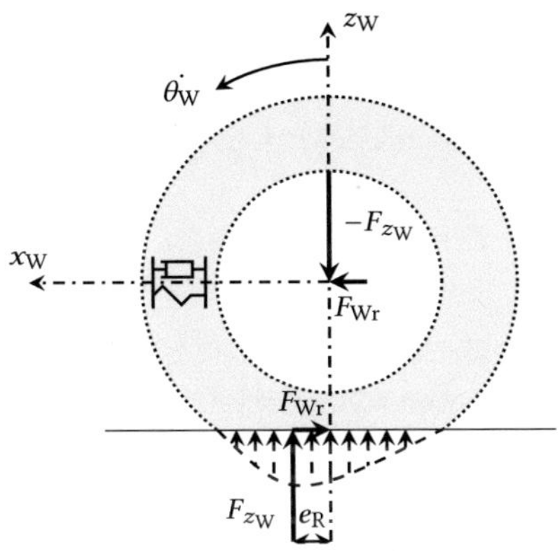

Figure 2.6: Tire analogous model [compare 155, p. 165].

The wheel vertical force (F_{z_W}) causes the compression of the tire in the area of the road surface and an area of tread deflection is developed as a contact surface. Usually a parallel-connected spring damper model is used to describe the compressible characteristics of the air inside the tire and damping characteristics of the rubber [69, pp. 137–138]. During the deflection at the tire inlet, spring and damper act simultaneously. During the rebound at the tire outlet the spring force is reduced by the damper. Thereby the surface pressure at the tread is imbalanced. The resulting force F_{z_W} is shifted from z_W about the distance e_R to the front. A torque appears which counteracts to the rolling direction. The ratio of the two lever arms e_R/r_W is called coefficient of rolling resistance (f_r) which is required to compute the rolling resistance

$$F_{Wr} = \frac{e_R}{r_W} F_{z_W} = f_r F_{z_W}. \tag{2.9}$$

Note that (2.9) is a simplified equation to compute the rolling resistance because the rotational part of the wheel air resistance was assumed negligible. The coefficient of rolling resistance depends on the wheel vertical force, tire temperature, internal tire pressure, vehicle velocity, and the road surface [69, p. 139]. While driving on paved roads, the progressive influence of the velocity on f_r is approximated in [114, p. 9] with

$$f_r(v_V) = f_{r0} + f_{r1} v_V + f_{r4} v_V^4. \tag{2.10}$$

In the lower speed range $v_V < 80\,\mathrm{km\,h^{-1}}$, $f_r(v_V)$ is almost linear [74, p. 41]. In conventional vehicles, the rolling resistance causes a purely dissipative heat which causes higher tire pressure and this heat is partly transferred from the tire into the environment.

2.4.5 Toe-in resistance

According to [69, p. 144] and [114, p. 14], the slackness within the steering is compensated by a light slant ($\alpha_W \approx \delta_W \ll 1°$) of the front wheels. By reason of this, a wheel lateral force develops which has a component contrary to the direction of movement. Accordingly, the toe-in resistance (F_{Wt}) is calculated by

$$F_{Wt}(\alpha_W) = F_{yW}(\alpha_W)\sin(\alpha_W).$$

2.4.6 Suspension resistance

While driving over a ground wave, a force originates which is normal and parallel to the road surface and transferred into the vehicle by springs and dampers of the suspension [69, p. 149]. Dampers show smaller compression than expansion rate and convert kinetic energy into thermal energy. Hence, the damper causes that this normal force is smaller during rebound then during compression. Hence, this normal force has a component contrary to the direction of movement and this component is called suspension resistance. In conventional vehicles, the suspension resistance causes dissipative heat transfer from the damper into the environment. However, recently there are approaches to gain energy out of the lifting movement of the chassis [84].

2.5 Lateral vehicle forces

The following forces act in vehicle lateral axis (y_V) and apply in vehicle center of gravity or in the pressure point. However, these forces may cause force components in x_V direction and hence influence the longitudinal vehicle forces as well.

2.5.1 Cornering resistance

Emerging on stationary cornering the centrifugal force (F_c) in (2.11) applies at the vehicle center of gravity and needs to be compensated by wheel lateral force (F_{yW}) at the wheels.

$$F_c = m_V \frac{v_V^2}{r} \tag{2.11}$$

Riekert and Schunck [137] introduced the *single track model* that is known as good approximation of lateral vehicle dynamics and that is shown in Figure 2.7.

The single track model makes the following assumptions:

1. constant vehicle velocity, $\dot{v}_V = 0$;
2. no motion in z_V direction, zero vehicle roll angle and vehicle pitch angle, and the vehicle center of gravity is assumed to lie within the road surface;
3. wheels are summarized to one wheel per axis;
4. the wheel vertical force balancing does not change over time,

[155, pp. 243,244]. The single track model is sufficiently exact for lateral accelerations which are smaller than $4\,\mathrm{m\,s^{-2}}$ [155, p. 244].

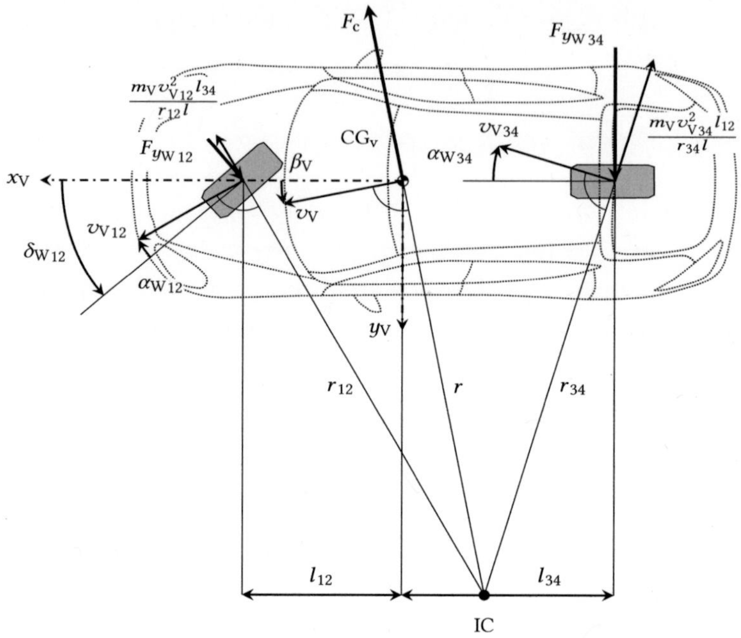

Figure 2.7: Vehicle single track model [compare 155, p. 245], [114, p. 553] and [69, p. 147]. The symbols denote: centrifugal force (F_c), wheel lateral force (F_{yW}), slip angle (α_W), wheel steer angle (δ_W), path radius (r), vehicle velocity (v_V), vehicle side slip angle (β_V), and wheel base (l).

Under the assumption of small slip angles the linear tire model (2.3) is used [69, p. 146] to get the relations

$$F_{yW\,12} = f(\alpha_{W12}) \approx X_{W12}\alpha_{W12} = \frac{m_V v_{V12}^2 l_{34}}{r_{12} l \cos \alpha_{W12}}, \tag{2.12a}$$

$$F_{yW\,34} = f(\alpha_{W34}) \approx X_{W34}\alpha_{W34} = \frac{m_V v_{V34}^2 l_{12}}{r_{34} l \cos \alpha_{W34}}, \tag{2.12b}$$

where l is the wheel base with $l = l_{12} + l_{34}$ and $X_{\mathrm{W}12}$, $X_{\mathrm{W}34}$ are the front and rear axle cornering stiffness, respectively. The relation (2.12) causes force components which act contrary to the direction of movement,

$$F_{\mathrm{Wc}12;34} = F_{y\mathrm{W}\,12;34} \sin \alpha_{\mathrm{W}\,12;34} \tag{2.13}$$

and these force components are summarized as cornering resistance (F_{Wc}). Combining (2.12) and (2.13) yields

$$\begin{aligned} F_{\mathrm{Wc}} = {} & \frac{m_{\mathrm{V}} v_{\mathrm{V}12}^2 l_{34}}{r_{12} l \cos \alpha_{\mathrm{W}12}} \sin\left(\frac{m_{\mathrm{V}} v_{\mathrm{V}12}^2 l_{34}}{r_{12} l \cos \alpha_{\mathrm{W}12} X_{12}}\right) + \dots \\ & \dots + \frac{m_{\mathrm{V}} v_{\mathrm{V}34}^2 l_{12}}{r_{34} l \cos \alpha_{\mathrm{W}34}} \sin\left(\frac{m_{\mathrm{V}} v_{\mathrm{V}34}^2 l_{12}}{r_{34} l \cos \alpha_{\mathrm{W}34} X_{34}}\right). \end{aligned}$$

Since there are only small slip angles while cornering, the angular relationships are simplified to $\cos \alpha_{\mathrm{W}} \approx 1$ and $\sin \alpha_{\mathrm{W}} \approx \alpha_{\mathrm{W}}$. Therefore the total cornering resistance is approximated by

$$F_{\mathrm{Wc}} \approx \frac{m_{\mathrm{V}}^2 l_{34}^2 v_{\mathrm{V}12}^4}{l^2 r^2{}_{12} X_{12}} + \frac{m_{\mathrm{V}}^2 l_{12}^2 v_{\mathrm{V}34}^4}{l^2 r^2{}_{34} X_{34}}, \tag{2.14}$$

[69, p. 147]. In the case of large curve radii, same set of tires, and the assumption that the vehicle center of gravity is in the middle of the vehicle, (2.14) can be further simplified into

$$F_{\mathrm{Wc}} \approx \frac{m_{\mathrm{V}}^2 v_{\mathrm{V}}^4}{r^2 4 c_{y\mathrm{W}}}, \tag{2.15}$$

[69, p. 147].

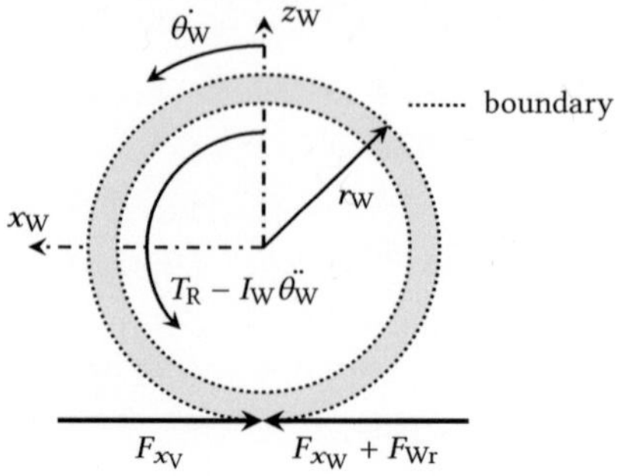

Figure 2.8: Forces in road surface and torque at a wheel that moves into x_W direction [compare 85, p. 37].

2.5.2 Lateral aerodynamic resistance

The lateral aerodynamic resistance (F_{y_a}) acts in pressure point and is similar to (2.7) given by

$$F_{y_a} = \frac{\rho_a}{2} A_V(\psi_a) v_a^2 c_y(\psi_a),$$

[69, p. 154]. Again, the lateral drag coefficient (c_y) is a function of ψ_a, [114, p. 62]. In contrast to longitudinal aerodynamic resistance, a torque raises around the vehicle vertical axis (z_V) from lateral aerodynamic resistance due to the lever arm between pressure point and vehicle center of gravity which is observable in Figure 2.5. This torque and the lateral aerodynamic resistance itself have to be compensated by wheel lateral force. Hence, the driver has to steer if $F_{y_a} > 0$ and accordingly, a resistance is introduced which is in principle comparable with the cornering resistance from Section 2.5.1. Mitschke and Wallentowitz [114, pp. 621–640] give a deeper introduction in vehicle side-wind dynamics.

2.6 Force equilibrium at wheel

Let us balance all forces in circumferential direction at the wheel by the free body diagram of the wheel in Figure 2.8. The force equilibrium becomes

$$F_{x\mathrm{V}} + F_{x\mathrm{W}} + F_{\mathrm{Wr}} = 0, \tag{2.16a}$$

$$F_{x\mathrm{V}} = \frac{T_{\mathrm{R}} - I_{\mathrm{W}}\ddot{\theta}_{\mathrm{W}}}{r_{\mathrm{W}}}. \tag{2.16b}$$

The vehicle tractive force ($F_{x\mathrm{V}}$) in (2.16b) includes the rim torque (T_{R}), the wheel rotation angle (θ_{W}), the wheel moment of inertia (I_{W}), and the dynamic wheel radius. The components of the rim torque will be explained in Section 2.7.

We will use (2.16) to derive two vehicle tractive force models in Section 5.4.2 and Section 5.4.3. Moreover, Section 1.1 said that the prediction of the vehicle tractive force is the main objective of this work and (2.16b) gives the expression of the vehicle tractive force.

2.7 Drive-train

The rim torque (T_{R}) is located at the internal section in Figure 2.8. Typically, it is expensive to measure T_{R} directly. Accordingly, vehicle specific *drive-train models* are used to determine T_{R} out of the engine torque (T_{E}) [see 91, pp. 194–221].

A drive-train model describes the lossy torque transmission from the engine to the rim. Losses arise through friction within the drive-train. Often the friction losses of single components such as bearings, gearbox, or differential are described with characteristic maps which originate from experiments. The exact setup of the drive-train model depends heavily on the used components (electric motor, hybrid, conventional combustion engine, two wheel drive, four wheel drive, manual gearbox, automatic gearbox, …).

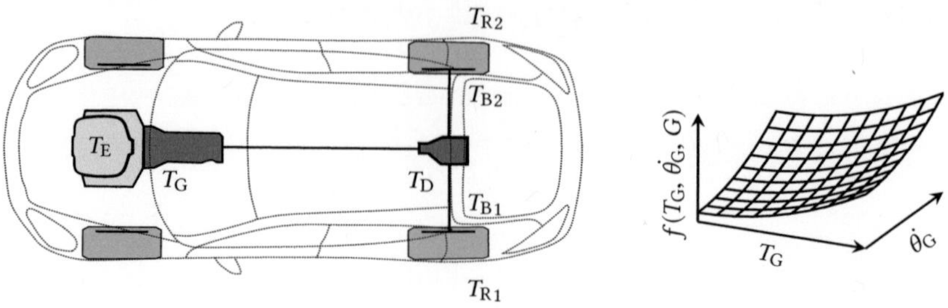

Figure 2.9: Drive-train of a two wheel drive vehicle with combustion engine and rear-drive. The right side shows the qualitative shape of the loss-torque as function of torque and rotational speed for one gear of the manual gearbox.

Therefore, it is impossible to find an uniform model structure for every thinkable drive-train and we will focus on one specific drive-train model. Figure 2.9 shows the drive-train of a two wheel drive vehicle with combustion engine and rear wheel drive. The engine torque is transformed in the gearbox and differential. We need to subtract the frictional losses which originate in this process at the gearbox output and differential output. While the clutch is engaged ($\dot{\theta}_G \approx \dot{\theta}_E$) the following relations give a drive-train model

$$\begin{aligned}
T_G &= T_E, \\
T_D &= T_G i_G(G) - f(T_G, \dot{\theta}_G, G), \\
\dot{\theta}_D &= \frac{\dot{\theta}_G}{i_G(G)}, \\
\sum_{jw=1}^{4} T_{R,jw} &= \Big(T_D i_D - f(T_D, \dot{\theta}_D)\Big) - \dots \\
&\qquad \dots - I_{red}(G)\frac{\ddot{\theta}_{W1} + \ddot{\theta}_{W2}}{2} - \sum_{jw=1}^{4} T_{B,jw}, \\
I_{red} &= \sum_{jw=1}^{4} I_{B,jw} + i_D^2 I_D + i_D^2 i_G^2(G)(I_G(G) + I_C + I_E).
\end{aligned} \tag{2.17}$$

T_G denotes the gearbox input torque, T_D the differential input torque, T_B denotes the braking torque, i_G the gearbox ratio, G the gear, $\dot{\theta_G}$ the derivative of the gear shaft rotation angle (gear shaft speed), $\dot{\theta_D}$ the derivative of the differential shaft rotation angle (differential shaft speed), i_D the differential ratio, I_C the clutch moment of inertia and I_E the engine moment of inertia. Here, it is assumed that all losses of auxiliary users (alternator, air conditioning compressor, ...) are already taken into account in T_E. Moreover, (2.17) is shown in a short form (not all required rotational inertias are explicitely shown). The extended form of (2.17) considers all rotational inertias within the drive-train and it refers to $\dot{\theta_W}$.
The wheel individual braking torque (T_{B,j_W}) is given by

$$T_{B,j_W} = p_{B,j_W} A_{B,j_W} r_{B,j_W} X_{B j_W},$$

[74, p. 171]. T_B is dissipative in conventional vehicles with disc brakes but may be zero while non-braking, $T_B \geq 0$.
X_B includes the braking friction coefficient between the brake pads and the disk as well as the brake caliper efficiency. The braking friction coefficient is subject to strong variation, depending on disk temperature and preconditioning of the brake [157, p. 32].
This work assumes, that T_R is available through a validated drive-train model for the used vehicle type. Thereby, all results can be transferred on similar vehicle types without further ado.

2.8 From equilibrium equations to models

Raising an equilibrium equation such as (2.16) is probably the standard method in engineering to deduce a model which approximates a mechanical system. A typical equilibrium is the *force equilibrium*

$$\sum_{i=1}^{j} F_i = 0. \tag{2.18}$$

Now assume that we consider three forces in (2.18), hence $j = 3$ than (2.18) becomes

$$F_1 + F_2 + F_3 = 0. \tag{2.19}$$

Let us suppose that F_3 is measurable. The vehicle tractive force from (2.16b) is for instance a measurable force if we use the drive-train model from Section 2.7. Let us further suppose that F_1, F_2 are measurable up to unknown parameters X_1, X_2. Then we can rewrite (2.19) into

$$\underbrace{A_{1,1}X_1}_{F_1} + \underbrace{A_{1,2}X_2}_{F_2} = \underbrace{B_1}_{-F_3}, \tag{2.20}$$

where A are the measured inputs, X are parameters and B is the measured output. We can substitute for example the acceleration force of (2.6) for F_1 in (2.20) with $A_{1,1} = \dot{v}_\mathrm{V}$ and $X_1 = m_\mathrm{V}$ and the longitudinal aerodynamic resistance (2.7) for F_2 in (2.20), while $A_{1,2}$ becomes v_V^2 and $X_2 = \frac{\rho_\mathrm{a}}{2} A_\mathrm{V} c_x$, assuming $\psi_\mathrm{a} = 0$.

Equation (2.20) is linear in the parameters and because of that, (2.20) is a linear multi-input-single-output (MISO) model which comprises only two parameters. Hence, the model structure of (2.20) is rather simple. We are free to choose a more flexible model structure if we increase the number of considered forces in (2.18) ($j > 3$). Accordingly, we can create a set of candidate models from the force equilibrium (2.18) with different complexity.

Commonly, X_1 and X_2 are only approximately known. Therefore, we try to estimate $\widehat{X}_1$ and $\widehat{X}_2$ with an overdetermined system of equations and $A_{1,1}$, $A_{1,2}$, and B_1, turn from scalars into the matrix form

$$\begin{bmatrix} A_{t=1,1} & A_{t=1,2} \\ A_{t=2,1} & A_{t=2,2} \\ \vdots & \vdots \\ A_{t=m,1} & A_{t=m,2} \end{bmatrix} \begin{bmatrix} X_1 \\ X_2 \end{bmatrix} \approx \begin{bmatrix} B_{t=1} \\ B_{t=2} \\ \vdots \\ B_{t=m} \end{bmatrix}. \tag{2.21}$$

Now the rows of A and B contain measurements depending on the time (t). Note the $\approx$ symbol in (2.21). This symbol indicates that we expect some uncertainty in the measurement A or B or in both of them. If we assume uncertainty solely in B, (2.21) can be written as multi-input-single-output output-error model with matrix notation as

$$B = A\widehat{X} + \Delta B,$$

whereas ΔB is the output correction (see Section 3.5). If we assume uncertainty in A and B, the multi-input-single-output errors-in-variables model of Section 3.6 holds

$$B = (A - \Delta A)\widehat{X} + \Delta B$$

and additionally considers the input correction (ΔA). Each assumption of uncertainty requires an individual estimator to determine $\widehat{X}$. Moreover, we can transfer (2.21) into a state-space output-error model

$$\begin{aligned} \widehat{X}_t &= \mathcal{A}\widehat{X}_{t-1} + \Delta X_t \\ B_t &= A_t\widehat{X}_t + \Delta B_t, \end{aligned}$$

where the state matrix ($\mathcal{A}$) considers knowledge about the temporal evolution of the parameters (see Section 3.7).

No matter what kind of model we use (multi-input-single-output, state-space, output-error, errors-in-variables), we will treat each row in (2.21) as independent measurement. Hence, we will not consider a specific structure in A and B.

However, most mechanical systems enforce a structure in A and B. Remember that we substituted the acceleration force for F_1 with $A_{1,1} = \dot{v}_V$ and $X_1 = m_V$. Hence, the measured vehicle acceleration ($\dot{v}_V$) over time gives the column vector $A_{:,1}$. A plot of $A_{:,1}$ would rather show a continuous function than a random signal, because the inertia of the vehicle causes that each row in $A_{:,1}$ correlates with adjacent rows. Accordingly, it might be useful to expand (2.21) in a way that A and X become

$$\left[\begin{array}{cc|c} A_{t=1,1} & A_{t=1,2} & \cdots \\ A_{t=2,1} & A_{t=2,2} & \cdots \\ \vdots & \vdots & \ddots \\ A_{t=m,1} & A_{t=m,2} & \cdots \end{array}\right] \left[\begin{array}{c} X_1 \\ X_2 \\ \hline \vdots \end{array}\right] \approx \begin{bmatrix} B_{t=1} \\ B_{t=2} \\ \vdots \\ B_{t=m} \end{bmatrix}. \tag{2.22}$$

The right block in the A-matrix of (2.22) could be filled with time-delayed measurements of some columns of A to consider the structure of A in the model. Moreover, we could fill this block with time-delayed measurements of B or ΔB to model the dynamics of unaccounted forces or to include a dynamical model of the output correction.

Models which combine a deterministic part (left block in the A-matrix of (2.22)) and a part for the disturbance (right block in the A-matrix of (2.22)) are called auto regressive with eXogenous input (ARX). Isermann and Münchhof [86, p. 57] provide an overview for this model structure. In addition to this, [96] compares various ARX models for the errors-in-variables problem. Apparently, the number of parameters in (2.22) is larger than in (2.21). Hence, the identification as well as the prediction of ARX models becomes computational more expensive, which is the main reason why we will not continue to consider ARX models.

On ✎ the other hand, this short introduction into ARX models might offer future research topics in vehicle parameter estimation.

Summary

This chapter described major vehicle force components that are required to form a force equilibrium which is fundamental to deduce vehicle models. We will introduce various models and estimators in Chapter 3, where statistical assumptions and relations between different models are emphasized. Further, the introduction in vehicle dynamics within this chapter is useful to follow Chapter 4 (Survey of related research). In Chapter 5, we will use the force equilibrium of (2.16) and perform the same steps from (2.18) to (2.21) to deduce vehicle tractive force models.

3 Models and estimators

> “Models [. . .] are only approximations to unknown reality or truth; there are no true models that perfectly reflect full reality.”
>
> *Burnham and Anderson [28, p. 264].*

Outline

This chapter begins with an overview of various models and estimators. Than, we will introduce general concepts of model selection and model validation, followed by important methods with regard to regularization and robustness. Moreover, we will discuss linear gray-box models and their estimators in detail and emphasize transitions from one estimator into another. Besides, we will introduce novel recursive estimators which will generalize (include as special case) and improve common estimators. Several reproducible experiments with increasing complexity will indicate which estimator should be used for a specific problem. Afterwards, a novel estimator for a polynomial-function black-box model will provide an improved signal filter which will be used in Chapter 5 to smooth vehicle CAN signals. Finally, we will outline additional topics which might motivate further research. We will conclude this chapter with a guide for estimator selection. The deeper study of linear gray-box estimators within this chapter is required to evaluate related research in Chapter 4 and to interpret the results of the real world problem in Chapter 5.

3.1 Fundamentals

Söderström and Stoica [168, p. 9] introduced the four concepts system, model structure, estimator and experimental condition which are useful to explain our degree of freedom with respect to system identification. The first concept *system* is the vehicle within this work. Usually, unmodified vehicles are desired. Hence, the system is given. In contrast, *models* are approximations of systems and we look often for models which describe the system response (output) for known input signals. Herein, we will focus on the vehicle longitudinal dynamics, precisely on models for the prediction of the vehicle tractive force. These models will be detailed in Section 5.4. Models can have different *model structures*, and some model structures have tunable parameters which can be adjusted through various estimators. An *estimator* is a mathematical procedure to determine unknown model parameters from measurements. The weather belongs to the *experimental condition*, which is hard to control. Practically, it is impossible to conduct two test rides under the same weather condition. Hence, we should assume that the experimental condition is given. However, we are free to choose an appropriate model structure and estimator.

Figure 3.1 gives an overview of various model structures and estimators. Starting from the root modeling, the three branches white-box, gray-box and black-box arise. We can differentiate these three branches by the amount of knowledge that each category requires to create models [86, p. 6].

First, *white-box* requires that everything about the system is known exactly to build a model. Therefore, we should be confident if we apply a white-box model. Commonly, white-box models arise from differential equations. Differential equations are based on physical laws, such as energy conservation laws, conservation of mass (fluid mechanics), or force-torque equilibrium, as discussed in Section 2.8. A simple example is the description of motion of a spring mass damping system using a differential equation with the exactly

known system-determining parameters mass, spring stiffness, and damping constant.
Second, *gray-box* requires physical insight in the system but offers some degree of freedom in choosing the system-determining parameters. To be more specific, using a gray-box model means that we are uncertain in the parameters. Mostly, gray-box models ground on physical laws and thus, on differential equations. Although by nature usually non-linear, many technical systems can be adequately approximated by linear models. From here, we can choose between multi-input-single-output (MISO), random-walk (RW), and state-space (STSP). If the approximation by linear models is no longer valid, more challenging non-linear gray-box models are required with basically the same subcategories random-walk and state-space.
Third, *black-box* means that we have no idea about the underlying physics of the system. Black-box models may be linear or non-linear and typical black-box models are impulse response (IR), polynomial-function (PF), neural networks (NN), and lookup tables. *Lookup tables* are often gained by measurements on test benches. The drive-train model of Section 2.7 for instance, is partly a black-box model with lookup tables for each gear. A lookup table is direct input-output mapping of data without the need to describe the data by mathematical equations. On the other hand, every time a lookup table is involved some kind of interpolation method is required for data which is not directly stored in the lookup table.
The explanation up to here was about different model structures. If we take a look further down of the gray-box or black-box branch in direction of the leaves, we discover that each model structure has various estimators. Herein, the *estimator* is nested under a model structure, considers statistical assumptions within a cost function, and provides an algorithm to estimate unknown parameters. Figure 3.2 visualizes the hierarchy of system, model with model structure, and estimator (batch or recursive algorithms) as general block diagram. Basically, the estimator adjusts parameters of a model with a given model structure. The estimator selects the model parameters in

such a way, that the difference between measured output (B) of the system and the model output gives the smallest possible value of the cost function.

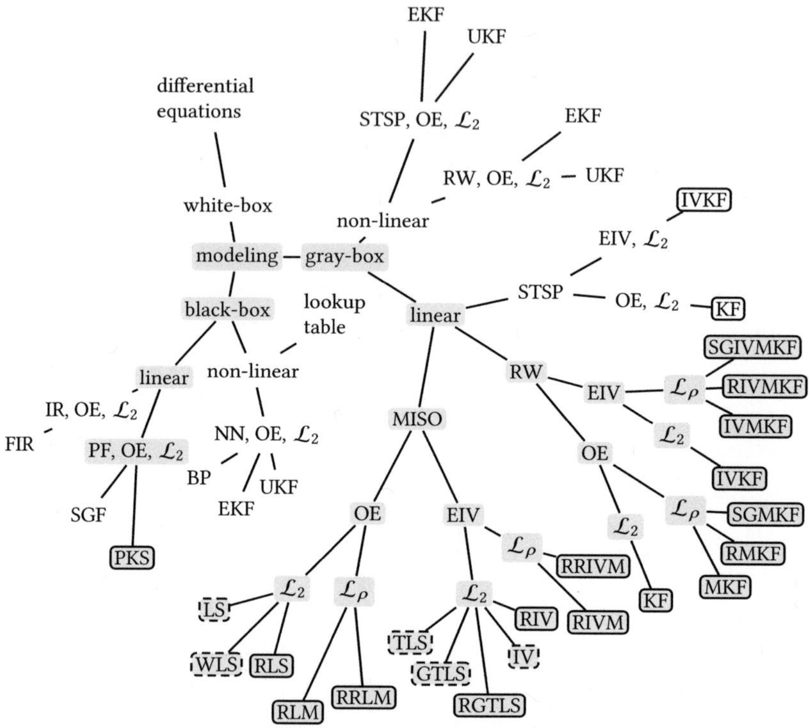

Figure 3.1: Model structures and estimators. [gray box] shows explained methods, [solid box] recursive algorithms and [dashed box] batch algorithms. $\mathcal{L}_2$ depicts non-robust, $\mathcal{L}_\rho$ robust estimators. The acronyms mean: backpropagation (BP), errors-in-variables (EIV), extended Kalman filter (EKF), finite impulse response (FIR), generalized total least squares (GTLS), impulse response (IR), instrumental variables (IV), IV Kalman filter (IVKF), IV M-Kalman filter (IVMKF), Kalman filter (KF), least squares (LS), multi-input-single-output (MISO), M-Kalman filter (MKF), neural networks (NN), output-error (OE), polynomial-function (PF), polynomial Kalman smoother (PKS), recursive GTLS (RGTLS), recursive IV (RIV), recursive IV M-estimator (RIVM), recursive M-estimator (RLM), recursive least squares (RLS), regularized M-Kalman filter (RMKF), recursive regularized IV M-estimator (RRIVM), regularized IV M-Kalman filter (RIVMKF), recursive regularized M-estimator (RRLM), random-walk (RW), Savitzky Golay filter (SGF), Stenlund-Gustafsson IV M-Kalman filter (SGIVMKF), Stenlund-Gustafsson M-Kalman filter (SGMKF), state-space (STSP), total least squares (TLS), unscented Kalman filter (UKF), weighted least squares (WLS).

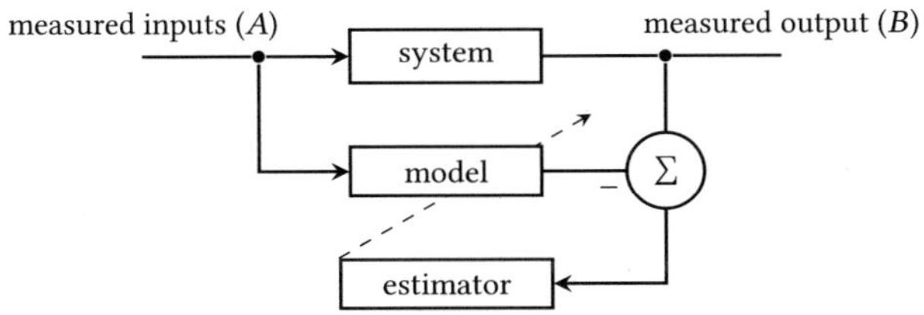

Figure 3.2: Block diagram of system, model (with a specific model structure), and estimator. Given the measurements, the estimator adjusts the model parameters, which is indicated by the dashed arrow. Disturbances are not shown [compare 45, pp. 3,8].

Hence, estimation here means mathematical optimization. For instance, least squares (LS) is a batch estimator for the gray-box linear multi-input-single-output output-error model structure with $\mathcal{L}_2$ cost function. Besides, there are recursive estimators which will be more important in the following for us.

There is a distinct difference between white-box on the one hand, and gray-box and black-box on the other. White-box models have not been adjusted by any kind of estimator, whereas gray-box and black-box models require estimators.

As Figure 3.1 comprises such a great diversity of model structures and estimators, it makes sense to introduce at first some methods to examine the *model quality*, which allows us to compare different models.

3.2 Model selection and model validation

Suppose we have a set of candidate models and want to compare these candidate models in terms of their model quality, then *model selection* gives the best model inside the examined set of candidate models [100, p. 509]. However, the set of candidate models might not contain a useful model at all. Hence, we need *model validation* to test if this best model is good enough for the intended purpose, if it explains the measured output well, and if it is able to describe the hidden true system [100, p. 509].

Before we introduce numerous model structures and estimators we need to define a goal. Our aim is to find a good model. The dictum on Page 31 already answers the question if we can find a perfect model. On the first look, this dictum is discouraging. However, we know now that there is no optimal model and we can focus on finding a good model. So, what is good? Ljung [100, p. 492] as well as Söderström and Stoica [168, pp. 423, 438] defined two principles: flexibility and parsimony. Flexibility means "Is the model structure large enough to cover the [...] system" [168, p. 423]? Parsimony means "Do not use extra parameters [...] if they are not needed" [168, p. 438]. To sum up, a good model is flexible enough (gives small bias) and parsimonious (gives small variance).

The *bias-variance dilemma* [73, p. 114] is a important concept in model selection. To explain the bias-variance dilemma, let us split a cost function ($\mathcal{L}$) into a sum of bias and variance [100, p. 492]

$$\mathcal{L} = \mathcal{L}_{\text{bias}} + \mathcal{L}_{\text{variance}}. \tag{3.1}$$

One should choose the model inside the set of candidate models with the smallest $\mathcal{L}_{\text{bias}}$ and $\mathcal{L}_{\text{variance}}$. However, in practice it turns out that small bias and small variance are hard to achieve simultaneously.

Haykin [73, p. 114] explains the bias term as the inability of the model to describe the physics of the system. Generally speaking, more complex model structures cover the physics of the system better and reduce the bias. However, complex model structures comprise usually plenty parameters and introduce a lot of uncertainty for the estimator, because the information in the training data may be not sufficient to identify these plenty parameters accurately. Hence, model selection is always a trade-off between bias (or flexibility) and variance (parsimony).

3.2.1 Error-based performance indices

We need measurable values (*performance indices*) to compare the model quality of different model structures and estimators. If the true parameters ($\overline{X}$) are known, we can compute the parameter error with the squared error vector norm (SEVN)

$$\mathrm{SEVN}_t = \left\| \overline{X}_t - \widehat{X}_t \right\|_2^2, \tag{3.2}$$

to observe which combination of model structure and estimator yields the best estimated parameters ($\widehat{X}$) [41, 53].

Common choices to measure the model quality, in terms of the *goodness of fit* between estimated output ($\widehat{B}$) and measured output (B), are the mean squared error (MSE), that gives an average over samples (m) in (3.3a) [100, p. 500], the normalized mean squared error (NMSE) in (3.3b) [100, p. 500] and the normalized root mean squared error (NRMSE) in (3.3c) [101, p. 8_15].

$$\mathrm{MSE} = \frac{1}{m} \left\| B - \widehat{B} \right\|_2^2 \tag{3.3a}$$

$$\mathrm{NMSE} = 1 - \frac{\left\| B - \widehat{B} \right\|_2^2}{\|B\|_2^2} \tag{3.3b}$$

$$\mathrm{NRMSE} = 1 - \frac{\left\| B - \widehat{B} \right\|_2}{\|B - \mu(B)\|_2} \tag{3.3c}$$

The two latter ones vary from bad goodness of fit ($-\infty$) to perfect goodness of fit (1) and measure how much of the measured output is explained by the model [100, p. 500].

3.2.2 Candidate models

How can we create a set of candidate models? First, we can define several candidate models from experience. Gray-box models require a certain knowledge of the underlying physics of the system. Hence, we can use this knowledge and setup some candidate models with increasing number of accounted forces for instance. In conclusion, the set of candidate gray-box models is commonly small and contains usually one useful model. However, we do not have much knowledge if we want to model a system with a black-box model. Hence, the second method is to create the set of candidate models rather randomly (trial and error).

There are more sophisticated methods to estimate a useful model structure such as rank tests of covariance matrices and canonical correlation [100, pp. 495–498]. The rank test of covariance matrices relies purely on data and yields parsimonious model structures without applying estimators. Canonical correlation is an iterative test if an additional parameter contributes to explain the measured output.

Additionally, *residual analysis* is a method to test if the selected model structure is flexible enough. If the residual analysis shows high correlation between output correction and past measured inputs in (3.4a), the model could be improved by adding one or more past measured inputs. The correlation between output correction and past output correction in (3.4b) should be small. Otherwise the output correction depends on past data and the model structure needs improvement [100, pp. 511–514].

$$R_{\Delta B,A;\tau} = \frac{1}{m} \sum_{t=1}^{m} \Delta B_t A_{t-\tau} \tag{3.4a}$$

$$R_{\Delta B;\tau} = \frac{1}{m} \sum_{t=1}^{m} \Delta B_t \Delta B_{t-\tau} \tag{3.4b}$$

3.2.3 Model selection with cross-validation

Let us start with a thought experiment to highlight the superior characteristics of cross-validation in model selection. Suppose we have recorded noisy data from a system which we want to model. This data is the *training data.*

Now we choose a suitable model structure and estimator. We identify a number of models with increasing model complexity (increasing number of parameters) from the training data and compute the goodness of fit for each model in terms of the MSE on the training data. It would turn out that the model with the highest model complexity (the most flexible model structure) yields the smallest MSE. As long as we increase the model complexity, the MSE will decrease further.

Now we are in a trap which is known as *overfit*. The models start to explain the noise with increasing model complexity [100, p. 501]. We could even achieve perfect fit with a very complex model. Overfitting is very dangerous. So, let us study how cross-validation works to avoid overfitting.

The simplest kind of *cross-validation* splits the recorded data into training data and *validation data* [100, p. 498]. Now we estimate the model parameters purely with the training data and compare the goodness of fit with the MSE for the validation data. Accordingly, we need to compute the estimated output for the validation data for each model and call this output the k step ahead prediction, where k is the number of prediction steps. As the validation data is unseen for each model, the MSE on the validation data shows the prediction quality for each model. Hence, cross-validation delivers unbiased estimates of the cost function in (3.1) $\mathbb{E}(\mathcal{L})$ [100, p. 501].

Note that model selection with cross-validation is a pragmatic method. No assumptions about the system or statistics are required and models obtained from different estimators are comparable. Coss-validation has one drawback. The training data is smaller due to the split of the recorded data into training data and validation data. However, sometimes this drawback can be overcome by applying the estimator for a second time on all data (training data and validation data) with the previously selected model structure.

3.2.4 Model selection with information criteria

Information criteria deliver the model quality without the need to portion the recorded data into training data and validation data. Hence, information criteria are the tools of choice if the recorded data from the system is expensive. However, information criteria rely on assumptions about statistics and vary with respect to the cost function of the estimator. Hence, it is not straight forward to compare different models which were gained by various estimators. While the derivation of the most common information criteria namely the Akaike's information criterion (AIC) is given in [100, pp. 501–504], let us focus more on practical aspects in applying the AIC. AIC in (3.5a) provides a measurable value which accounts simultaneously for model quality and model complexity [27, p. 61]. The first term in (3.5a) measures model quality through the value of the cost function ($\mathcal{L}$) and the second term the model complexity through the number of estimable parameters (o) which comprises the model parameters as well as parameters of the assumed statistical distribution. Note that AIC is given in (3.5a) without specific statistical assumptions. Equation (3.5b) gives AIC with the cost function for the least squares estimator [168, p. 442].

$$\mathrm{AIC} = -2\log\mathcal{L} + 2o \tag{3.5a}$$

$$\mathrm{AIC} = m\log\left(\frac{1}{m}\sum_{i=1}^{m}\left(B_i - \widehat{B}_i\right)^2\right) + 2o \tag{3.5b}$$

Generally, the smaller the AIC, the better the model. It is not the goal to find a model which memorizes the data, but to find a model that learns and captures the hidden information in the data [28, p. 275]. The pure AIC values are not directly applicable as they depend strong on the samples (m). According to [27, p. 70], the AIC difference (ΔAIC) is more meaningful to decide which model describes the hidden information superiorly.

$$\Delta\mathrm{AIC}_i = \mathrm{AIC}_i - \min(\mathrm{AIC})$$

Table 3.1: Recommendation to evaluate the respectively model quality with $\Delta\mathrm{AIC}_i$ [reproduced from 27, p. 70].

$\Delta\mathrm{AIC}_i$	support of model
0 to 2	substancial
4 to 7	considerably less
>10	essentially none

The best model from the set of candidate models obtains a value of $\Delta\mathrm{AIC}_i = 0$. Table 3.1 gives recommendations how the respective $\Delta\mathrm{AIC}_i$ are evaluated. Akaike's information criterion can only compare the results of several models on the same record [27, p. 334].
Another often applied information criteria is the Bayesian information criterion (BIC) [27, p. 286] in (3.6a). The BIC gives the best model for prediction and grounds opposed to AIC on Bayesian theory. In (3.6b), BIC is given with the cost function for the least squares estimator.

$$\mathrm{BIC} = -2\log\mathcal{L} + o\log m \tag{3.6a}$$

$$\mathrm{BIC} = m\log\left(\frac{1}{m}\sum_{i=1}^{m}\left(B_i - \widehat{B}_i\right)^2\right) + o\log m \tag{3.6b}$$

The difference between AIC and BIC is explained in [27, pp. 293–301]. Basically, BIC compares the model quality purely with Bayesian theory and makes no assumptions about the true system whereas AIC tries to find the model which is close to the true system and makes assumptions about the system. The true model is not required in the set of models in the derivation of BIC [27, p. 295], whereas the derivation of AIC believes that a perfect model exists which is equal to the true system. Which information criteria is to be preferred depends strongly on the problem and there is no general rule if AIC or BIC is the better choice.

3.2.5 Model validation

Once we found the best model inside the set of candidate models via model selection, model validation provides confidence that this model works for the intended purpose. Remember, it might be that the set of candidate models does not include a useful model at all.

Ljung [100, p. 509] highlights that the ultimate model validation is the successful application of the model. Basically, if the model works satisfactory all theoretic concerns about statistics, the model structure, and so forth are negligible. However, it is usually expensive and sometimes dangerous to validate a model directly at the intended application.

Cross-validation from Section 3.2.3 is the superior method to validate a model [100, p. 510]. Accordingly, model selection with cross-validation yields the best and valid model from the set of candidate models in one step. Moreover, we established that gray-box models have parameters with physical meaning. Hence, if the parameter values, parameter variance, and the input-output sensitivity of the model confirms with prior knowledge of the system, convincing arguments are found that the model is reasonable [100, p. 509].

Before we begin to analyze the various model structures of Figure 3.1, we need to introduce two additional and important methods: regularization, and robustness.

3.3 Regularization

Let us suppose we want to solve the overdetermined linear problem

$$AX \approx B \tag{3.7}$$

where $A \in \mathbb{R}^{m \times n}$ is the measured input, $X \in \mathbb{R}^{n \times d}$ are the unknown parameters and $B \in \mathbb{R}^{m \times d}$ is the measured output. Problem (3.7) is an inverse problem, which is often ill-posed. *Ill-posed* means that the solution for X is sensitive to small variations in A and B [105], [160, p. 1].

Golub and Van Loan [63, pp. 80–81] explain that the measure of sensitivity can be evaluated by the condition number (c) of A. Let $USV = \text{svd}(A)$ be the singular value decomposition ($\text{svd}(\cdot)$) of A, then the condition number becomes $c = S_{1,1}/S_{n,n}$, where S is the matrix of eigenvalues (S). The condition number is within range $1 \leq c \leq \infty$ and the best possible condition number is one. If $c \gg 1$, problem (3.7) is *ill-conditioned* and the solution for X becomes unstable and may become meaningless [95, p. 1].
A problem is *well-posed* if Hadamard's conditions Existence, Uniqueness and Continuity hold [73, p. 343]. Following the discussion of [73, p. 343], we can conclude that Hadamard's conditions cannot be ensured if we apply recursive estimators.

> First, the existence criterion may be violated in that a distinct output may not exist for every input. Second, there may be not as much information in the training sample as we really need for a unique reconstruction of the input-output mapping [...]. Third, the unavoidable presence of noise [...] adds uncertainty to the reconstruction process [73, p. 343].

However, we will prefer recursive estimators instead of batch estimators, because batch estimators cause much computational burden.
Practically speaking, problem (3.7) may become ill-conditioned or singular due to poor excitation of the measured inputs that is known as *wind-up* problem [68, 186]. One or more measured inputs show poor excitation if the persistent excitation condition

$$\rho_1 I > \sum_{i=t}^{t+j} A_i^\top A_i > \rho_2 I$$

($\rho_1, \rho_2 > 0$) is not fulfilled [10, p. 136], [86, p. 250].

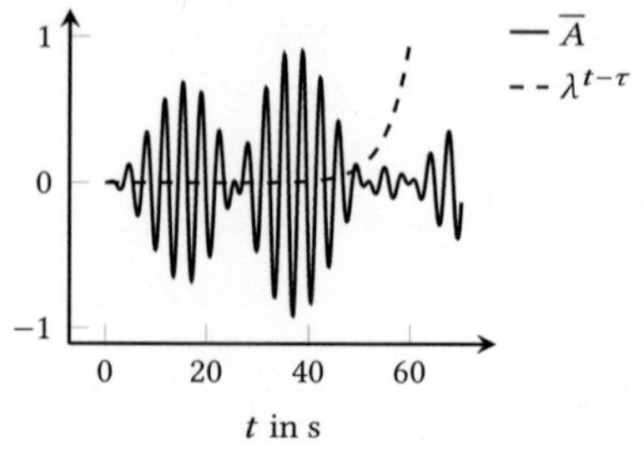

Figure 3.3: The true input $\overline{A}$ shows a mixture of *rich excitation* (10 s to 50 s) and *poor excitation* (50 s to 65 s). *Exponential forgetting* with the forgetting factor (λ) gives high weight to data where $\overline{A}$ shows poor excitation. Standard recursive estimators may suffer in this case under wind-up.

In short, poor excitation violates Hadamard's existence criterion, because the input-output mapping contains only few information.
Figure 3.3 provides an example where the true input ($\overline{A}$) shows poor excitation between 50 s to 65 s. If A is noisy, the observations might contain pure noise and the solution for X becomes uncertain and unstable if exponential forgetting is applied, which is common in recursive estimators. Several approaches have been introduced to avoid wind-up [30, 50, 124, 145, 146, 171]. *Regularization* is a mathematical method mainly introduced through Tikhonov [173] to stabilize (smooth) the solution of ill-posed problems. The basic idea of *Tikhonov regularization*, which is also known as ridge regression, is to augment the cost function with a regularizer as shown in (3.8a).

$$\min_X \mathcal{L} + \kappa ||\mathcal{R}X||_2^2 \tag{3.8a}$$

$$\min_X ||B - AX||_2^2 + \kappa ||\mathcal{R}X||_2^2 \tag{3.8b}$$

Substituting $\mathcal{L}$ with the LS cost function in (3.8b) yields the basic regularized LS cost function with the regularization parameter (κ) and the regularization matrix ($\mathcal{R}$) which is often set to the identity matrix (I) [160, p. 7]. Other types

of $\mathcal{R}$ are extensively discussed in [95, pp. 12–21] and shortly introduced in [160, p. 7].
If $\kappa = 0$, (3.8b) turns into the basic LS cost function. The left hand side in (3.8a) can be replaced with other cost functions such as the total least squares (TLS) cost function as shown in [95, p. 1], [160, p. 45], [105], [182, pp. 57–66], and [103]. The larger κ, the smoother the solution for X becomes. However, a too large κ reduces the influence of $\mathcal{L}$ in (3.8a). Hence, the choice of κ is always a trade-off between a cost function and the regularizer (smoothness of X). Often, we need to choose κ empirically. However, methods to adjust the optimal κ such as *L-curve* or cross-validation are explained in [160, pp. 11–13], [73, pp. 364–370], [34], [95, pp. 49–57], and [103].
Beside Tikhonov regularization other famous regularization methods are *Levenberg-Marquardt regularization* and *Lasso*, detailed in [186] and [172], respectively. Lasso favors sparse solutions through replacing the squared Euclidean norm by the 1-norm in the regularizer of (3.8b). Hence, Lasso with LS cost function becomes

$$\min_X \|B - AX\|_2^2 + \kappa \|\mathcal{R}X\|_1 .$$

3.4 Robust estimators

Most estimators are based on statistical assumptions of distributions. LS is probably the most applied and studied estimator. Rousseeuw and Leroy [142, p. 2] explain the outstanding popularity of LS until nowadays with two facts. First, LS has a closed-form solution which requires only simple matrix algebra. Second, LS yields the arithmetic mean of one dimensional data, which was around 1800 the most reliable location estimator. Notice that Gauss introduced the Gaussian distribution after defining the LS cost function so that LS is optimal for Gaussian distributed, zero mean output corrections [142, p. 2].

The LS cost function is

$$\min_X \sum_{i=1}^{m} (B_i - A_i X)^2 \tag{3.9}$$

[82, p. 155]. Note that other references give (3.9) with $\sum 1/2(\cdot)$, but this equation leads to the same solution.
The references given by [142, p. 3] and [98] show, that many real world output corrections are non-Gaussian distributed. For instance, Clancey [38] examined around 250 distributions of chemical analysis and only 10 % to 15 % could be treated as Gaussian distributed. Zoubir et al. [200, p. 62] provide references where impulsive noise (Gaussian like distributions with heavier tails) was found in signal processing problems.
But what happens if the output corrections are non-Gaussian? Many references show that LS fails to produce meaningful results if a single outlier is in the training data. Rousseeuw and Leroy [142, pp. 4,5] provide two examples of a straight line fit where a single outlier leads to corrupted parameters. A similar example is given by [200, p. 65].
Outliers can occur in the measured inputs (A) or in the measured outputs (B) [142, p. 5]. Applying the LS cost function, outliers in the measured inputs cause larger output corrections than outliers in the measured outputs. Outliers in the measured inputs are sometimes leverage points. In that case, they have a large influence on the LS solution. Rousseeuw and Leroy [142, p. 6] define a *leverage point* as a single point A_i, that lies far away from the majority of A. Hence, leverage points are not necessary outliers, but in any case they determine the LS solution strongly.

Experiment 3.1
Call the function outliersLeveragePoints() with the string 'LS' to compute the four LS straight line fits of Figure 3.4.

Figure 3.4 shows four straight line fits of seven data points with Gaussian output noise ($\widetilde{B}$). Each line fit is the result of LS estimation of the slope parameter, which has a true value of $\overline{X} = 3$. Figure 3.4a does not contain outliers in the data. Hence, the LS solution in Figure 3.4a is precise and near the true parameter. The accuracy of LS decreases significantly in Figure 3.4b with a single outlier in B_2 (the second entry of B). The original position of B_2 from Figure 3.4a is shown with an asterisk for convenience. The leverage point A_4 (the fourth entry of A) in Figure 3.4c is an outlier and hence, the LS solution becomes meaningless. However, a good leverage point does not corrupt LS as shown in Figure 3.4d, where the seventh data point was moved far away from the other six data points.

It was quite simple to find the leverage point in Figure 3.4c and Figure 3.4d. However, Rousseeuw and Leroy [142, p. 7] explain that it is impossible to uncover leverage points (and thus possible outliers) through simple visualization and inspection of the training data in higher dimensions. *Outlier diagnostics* provides data-based methods (no model is required) to uncover and remove outliers. Liang and Kvalheim [98] and Rousseeuw and Leroy [142, pp. 216–237] provide an overview of classic outlier diagnostics with a discussion of their drawbacks (uncovered leverage points), whereas Knorr and Ng [93] and Rousseeuw and Leroy [142, pp. 237–245] introduce more recent and improved methods. The surveys [36] and [76] provide extensive introduction into outlier diagnostics.

There is no general definition for outliers. Herein:

Definition 3.1. An *outlier* is an observation that somehow deviates from the assumptions, mainly assumptions about the distribution.

This definition matches well with the term *robust.* Huber and Ronchetti define robustness with: “robustness signifies insensitivity to small deviations from the assumptions” [82, p. 2]. Hence, robust estimators can deal with deviations from the assumptions or in other words can deal with a certain fraction of outliers.

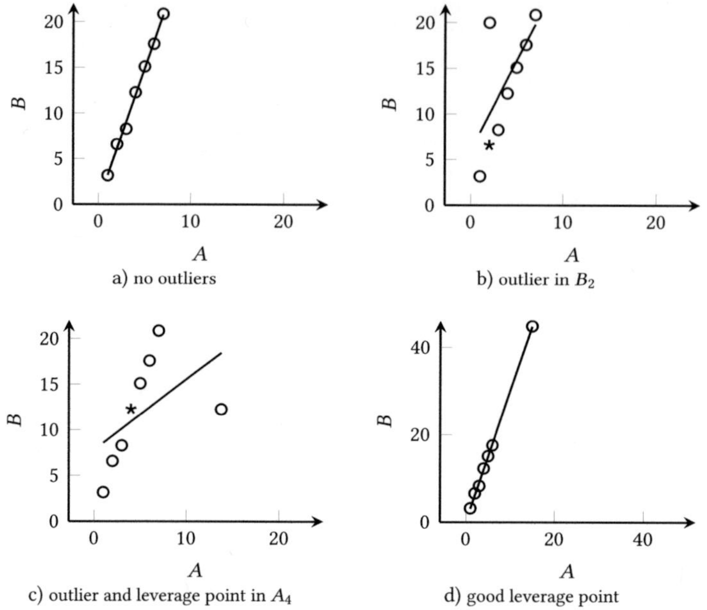

Figure 3.4: LS estimation (——) of seven data points (○) without outliers in Figure 3.4a, an outlier in the second entry of the measured output B_2 in Figure 3.4b, an outlier in the measured input A_4 in Figure 3.4c which is also a leverage point and a good leverage point in Figure 3.4d. The asterisk ($\star$) shows the original data A_4, B_2 of Figure 3.4a. Similar figures can be found in [142, pp. 4–6] and [200, p. 65].

The *breakdown point* is a concept to measure robustness. Rousseeuw and Leroy [142, p. 10] explain the breakdown point as smallest fraction of outliers that cause the estimator to produce solutions that are arbitrarily far from the consistent solution. In accordance Huber and Ronchetti define the breakdown point as:

> The breakdown point is the smallest fraction of bad observations that may cause an estimator to take on arbitrarily large aberrant values [82, p. 8].

LS is non-robust and has a breakdown point of 0 %, whereas the highest possible breakdown point is 50 % [142, pp. 15, 126].

3.4.1 High breakdown point methods

Huber and Ronchetti [82, p. 197] split robust estimators into robust methods and high breakdown point methods. High breakdown point methods seek to achieve the maximum breakdown point of 50 % but require often extensive computation or show a poor convergence. The convergence rate of LS is high with $m^{-\frac{1}{2}}$, if ΔB is Gaussian [82, p. 196].
Rousseeuw [139] introduced least median of squares (LMS) as high breakdown point estimator which can deal with outliers in the measured inputs and measured outputs. Therefore, LMS is robust in terms of bad leverage points (outliers in the measured inputs). The key idea in LMS is to replace the sum in the LS cost function (3.9) with the highly robust median (med(·)). Hence, the LMS cost function becomes:

$$\min_{X} \operatorname*{med}_{i} (B_i - A_i X)^2. \tag{3.10}$$

The solution of (3.10) is expensive in matters of computation. Subsamples (tuples) of the training data are required as described in [142, pp. 197–204], where the LMS algorithm is shown. Furthermore, the convergence rate is poor with $m^{-\frac{1}{3}}$ [82, p. 196].
Rousseeuw and Leroy [142, p. 15] introduced least trimmed squares (LTS) to overcome the problem with the poor convergence rate. LTS is as robust as LMS and has the same high convergence rate as LS. Rousseeuw and Leroy [142, p. 15] replaced the sum in the LS cost function with a trimmed sum:

$$\min_{X} \sum_{i=1}^{j} (\Delta B_i)^2, \tag{3.11}$$

where $\Delta B_i = B_i - A_i \widehat{X}$ are the (first squared and than sorted) output corrections and j is the last considered sample from all m samples. Note that the LTS cost function (3.11) equals the LS cost function (3.9) if $j = m$. Hence, the largest output corrections are not considered in (3.11), which makes

LTS robust but the additional sorting of the output corrections is even more pricy than the computation of the median in LMS. Rousseeuw and Leroy [142, pp. 206–208] outlined the root LTS algorithm with high computational complexity.

However, the favorite properties of LTS, namely high convergence rate, a breakdown point of 50 % and robustness against outliers in measured inputs and measured outputs motivated numerous researchers to reduce the required computational burden [2, 14, 77, 144].

In spite of these enhancements, to best of my knowledge it has not been shown how to apply LMS or LTS as recursive estimator. There is one major reason why it is unlikely that LMS or LTS may be used recursively. LMS and LTS require to split the training data into tuples. On the contrary, recursive estimators must deal with an indefinite growth of measurements ($m \longrightarrow \infty$). How should we create all possible (or at least some) tuples if the training data is consistently growing? This is the reason why Figure 3.1 does not show LMS and LTS and why we focus on M-estimators from now on.

3.4.2 M-estimators

Huber [81] introduced *M-estimators* in 1964 with the idea to change the squares in the LS cost function with another function. Therefore, the M-estimator cost function looks similar like the LS cost function and becomes

$$\min_X \sum_{i=1}^{m} \rho(B_i - A_i X), \tag{3.12}$$

where ρ is a symmetric function preferably with a unique minimum at zero. LS is a special type of M-estimator. If ρ is set to $\rho = (\cdot)^2$ (or $1/2(\cdot)^2$, see the note on (3.9)), (3.12) turns into (3.9). Hence, the proper choice of the ρ-function determines the robustness of M-estimators. Taking the derivative of (3.12) with respect to X yields the set of equations

$$\sum_{i=1}^{m} \psi\left(\frac{B_i - A_i X}{\widehat{\sigma}(\Delta B)}\right) A_{i,j} = 0, \quad j = 1, \ldots, n, \tag{3.13}$$

where ψ is the derivative of ρ with respect to ΔB ($\psi = \mathrm{d}\rho/\mathrm{d}\Delta B$) and $\widehat{\sigma}$ is the inserted estimated scale that standardizes the output corrections [82, p. 161]. The estimated scale is needed because M-estimators are not scale-invariant. Accordingly, the estimated scale needs to be estimated simultaneously by a robust estimator and the common choice is the median absolute deviation (MAD) [141, 200]

$$\widehat{\sigma}_{\mathrm{MAD}}(\Delta B) = 1.483 \operatorname{med} |\Delta B - \operatorname{med} \Delta B|, \tag{3.14}$$

where the factor 1.483 was introduced to achieve a consistent estimator of the standard deviation for the Gaussian distribution.
The cost function (3.13) requires iterative procedures because ρ and ψ are functions of the output corrections (ΔB) and therefore of the desired parameters (X). Iteratively weighted least squares (WLS) is the method of choice to solve (3.13) in practice, where the scaling matrix (W) has to be computed at each iteration with

$$W_{i,i} = \frac{\psi_i}{\Delta B_i / \widehat{\sigma}(\Delta B)}$$

until the estimated parameters remain sufficiently close together between the last and the current iteration. Algorithms of M-estimators, which are solved by iteratively WLS, are outlined in [82, p. 179], [98] and [21, p. 105]. We will use $\widehat{\sigma}$ as shorthand notation for $\widehat{\sigma}(\Delta B)$ in the following. Huber and Ronchetti [82, p. 174] suggest to use the following ρ-function and influence function

$$\rho(\Delta B/\widehat{\sigma}) = \begin{cases} \frac{1}{2}(\Delta B/\widehat{\sigma})^2 & \text{for } |\Delta B/\widehat{\sigma}| < \delta, \\ \delta\left(|\Delta B/\widehat{\sigma}| - \frac{\delta}{2}\right) & \text{for } |\Delta B/\widehat{\sigma}| \geq \delta, \end{cases} \tag{3.15a}$$

$$\psi(\Delta B/\widehat{\sigma}) = \begin{cases} \Delta B/\widehat{\sigma} & \text{for } |\Delta B/\widehat{\sigma}| < \delta, \\ \delta \operatorname{sgn}(\Delta B/\widehat{\sigma}) & \text{for } |\Delta B/\widehat{\sigma}| \geq \delta, \end{cases} \tag{3.15b}$$

where ψ is saturated to δ if $|\Delta B/\widehat{\sigma}| \geq \delta$. Contrary, ρ and ψ become for LS

$$\rho(\Delta B/\widehat{\sigma}) = \frac{1}{2}(\Delta B/\widehat{\sigma})^2, \tag{3.16a}$$

$$\psi(\Delta B/\widehat{\sigma}) = \Delta B/\widehat{\sigma}. \tag{3.16b}$$

We can see from (3.16b) that the influence function of LS is unbounded. Thus, LS is not robust if unusual large output corrections occur.

Brabanter et al. [21, pp. 100–110] propose to use the highly robust Myriad function to deal with extreme outliers. However, the convergence rate is lower than for the Huber functions (3.15). Myriad was designed as maximum Likelihood scale estimator for the Cauchy distribution which is similar to the Gaussian distribution, but has heavy tails. Therefore, Myriad is a good choice in impulsive noise environments [see 64]. The Myriad ρ and ψ functions are

$$\rho(\Delta B/\widehat{\sigma}) = \frac{\ln(10)}{2}\beta^2\Big(\log\big(\beta^2 + (\Delta B/\widehat{\sigma})^2\big) - \log\big(\beta^2\big)\Big), \tag{3.17a}$$

$$\psi(\Delta B/\widehat{\sigma}) = \frac{\beta^2 \Delta B/\widehat{\sigma}}{\beta^2 + (\Delta B/\widehat{\sigma})^2}. \tag{3.17b}$$

Rather than [21, p. 105], I added in equation (3.17a) additional scaling terms to achieve a unique minimum at zero. Yet, these scaling terms are more of theoretical interest, because the derivative of (3.17a) leads to the same influence function (3.17b) as used in [21, p. 105]. Other ρ-functions are shown in [82, p. 99], [21, p. 105] and [158].

Figure 3.5 visualizes the properties of LS, Huber, and Myriad. First, the ρ-functions in Figure 3.5a show the transition from LS cost function into Huber cost function. If the output correction is larger than δ, the quadratic criterion turns into a linear, compare (3.16a) with (3.15a). Second, we can observe Huber's intention directly in Figure 3.5b, where the influence functions are shown. LS has an unbounded influence function, whereas the Huber influence function is limited if the output correction is larger than δ. Myriad goes even further and gives gross output corrections decreasing influence. Third, Figure 3.5c gives the weight which is needed to fill the scaling matrix of iteratively WLS to solve (3.13). LS gives the same weight to all output corrections, Huber reduces the weight if $|\Delta B| \geq \delta$ and Myriad shows a weight-function which recalls the standard Cauchy distribution $f(x; 0, 1) = \frac{1}{\pi(1+x^2)}$.

There is one drawback of M-estimators compared with the LMS and LTS high breakdown point estimators. M-estimators are not robust in terms of bad leverage points. Actually, this fact motivated Rousseeuw to develop LMS and LTS although generalized M-estimators were introduced to improve the robustness of M-estimators [for further detail consult 139, pp. 12–14]. Cases where M-estimators fail because of one or more bad leverage points are shown in [139, 200]. Accordingly, the breakdown point of M-estimators is 0 %.

Figure 3.6 is the robust counterpart of Figure 3.4. The same data was used. The Huber tuning constant was $\delta = 1.345$ and the Myriad tuning constant $\beta = 1$. All robust estimators produce close results in Figure 3.6a and in Figure 3.6b, where LS showed a corrupted result (Figure 3.4b). Figure 3.6c shows the difference between a high breakdown point estimator (LMS) and a robust estimator (M-estimator). Both M-estimators perform not better than LS in Figure 3.4c. Yet, the Myriad M-estimator is slightly better than the Huber M-estimator. However, if the leverage point moves slightly more to the right, Myriad and Huber give close unsatisfactory results, whereas the performance of LMS remains superior.

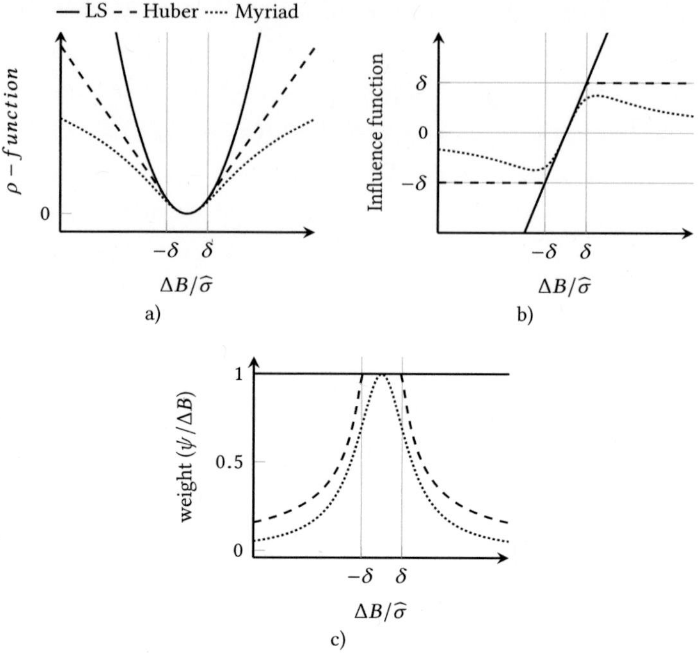

Figure 3.5: Various $\rho - functions$ in Figure 3.5a, influence functions in Figure 3.5b and the weight in Figure 3.5c.

Experiment 3.2

Call the function outliersLeveragePoints() with the string 'robust' to compute the four robust straight line fits of Figure 3.6. Modify the position of the bad leverage point in line 20 of outliersLeveragePoints() to bring the Myriad M-estimator to its breakdown.

On the other hand, we have seen in Figure 3.6b that M-estimators are highly robust against outliers in the measured output. Furthermore, Huber and Ronchetti remark:

> It appears that M-estimates offer enough flexibility and are by far the easiest to cope with, simultaneously, with regard to computation, asymptotic theory, and intuitive interpretation; moreover, the step from [...] [(3.9) to (3.12)] is easily explainable to nonstatisticians also [82, p. 164].

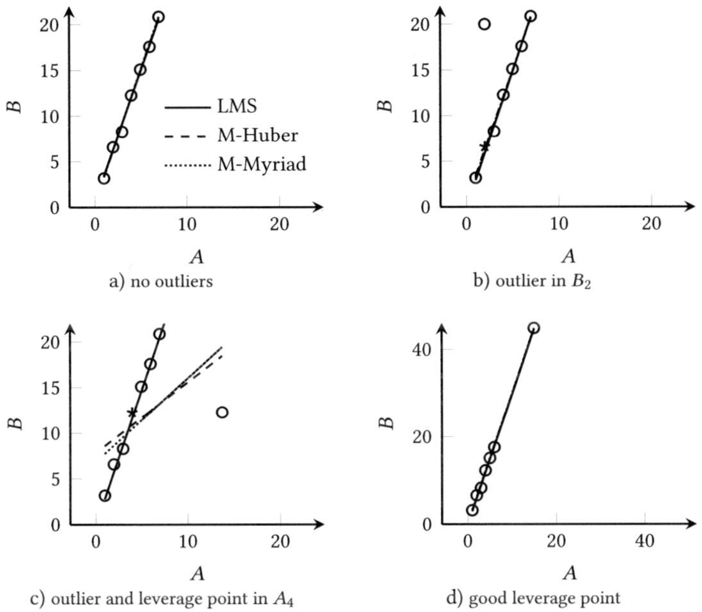

Figure 3.6: Robust estimation of the same data as in Figure 3.4. All estimators are close in Figure 3.6a and Figure 3.6d. Figure 3.6b shows the benefit of robust estimators compared with Figure 3.4b where LS failed. The outlier and leverage point in Figure 3.6c leads to a breakdown of the M-estimators. LMS is superior.

Let us add to this convincing comments the feature that M-estimators are an ideal source to develop recursive estimators. That is due to the relation between (iteratively) WLS and recursive least squares (RLS), where the latter will be the first presented recursive estimator in Section 3.5.3. Moreover, M-estimators are a vital part of other popular high breakdown point methods such as the MM-estimator [195] which performs superior in some signal processing applications as discussed in [200].

3.5 Linear multi-input-single-output output-error model

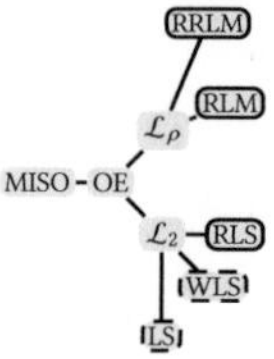

The first presented gray-box model is the linear multi-input-single-output output-error model. The respective branch from Figure 3.1 is shown in the margin for convenience. The model structure was already introduced in (3.7) and is detailed here with output corrections

$$AX \approx B, \quad B = \overline{B} + \widetilde{B}, \tag{3.18a}$$

$$B = A\widehat{X} + \Delta B, \tag{3.18b}$$

where the output correction is $\Delta B = B - \widehat{B}$ with $\Delta B \in \mathbb{R}^{m \times d}$. The number of outputs (d) is one, because we consider the single-output case ($d = 1$). Model structure (3.18) considers perturbation only in B (output-error). Hence, (3.18) is a constrained perturbation problem. No corrections are applied in A. Therefore, the noise covariance matrix ($\widetilde{P}$) is assumed with

$$\widetilde{P} = \operatorname{cov}\left(\begin{bmatrix}\widetilde{A} & \widetilde{B}\end{bmatrix}\right) = \sigma^2 \operatorname{diag}\left(\begin{bmatrix}0 & 0 & \dots & 1\end{bmatrix}^\top\right), \tag{3.19}$$

where $\operatorname{cov}(\cdot)$ is the covariance operator and $\operatorname{diag}(\cdot)$ extracts diagonal elements of a matrix or converts a vector into a diagonal matrix.

3.5.1 Least squares

The well known LS closed-form solution for problem (3.18) becomes

$$\widehat{X} = \left(A^\top A\right)^{-1} A^\top B \tag{3.20}$$

[37, p. 4]. The covariance of parameter estimation error $\mathbb{E}\left((\overline{X} - \widehat{X})(\overline{X} - \widehat{X})^\top\right)$ (denoted as covariance matrix (P) in the sequel) is commonly computed with $P = \widehat{\sigma}^2 (A^\top A)^{-1}$ and $\widehat{\sigma}^2 \approx \frac{\Delta B^\top \Delta B}{m-n}$ [168, pp. 65–66].

Proof. We can rewrite the LS cost function (3.9) in matrix notation

$$\min_X \|B - AX\|_2^2 \tag{3.21}$$

and follow Söderström and Stoica [168, pp. 62–63] to present (3.21) in the equivalent form

$$\begin{aligned}\mathcal{L} &= (B - AX)^\top (B - AX),\\ &= \left(B^\top - X^\top A^\top\right)(B - AX),\\ &= B^\top B - B^\top AX - X^\top A^\top B + X^\top A^\top AX.\end{aligned}$$

The following intermediate step explains how to compute the derivative for the rightmost term of the cost function.

$$\begin{aligned}\frac{\mathrm{d}}{\mathrm{d}X} X^\top A^\top AX &= \frac{\mathrm{d}}{\mathrm{d}X^*} X^{*\top} A^{*\top} A'X' + \frac{\mathrm{d}}{\mathrm{d}X'} X^{*\top} A^{*\top} A'X',\\ &= A^{*\top} A'X' + \left(X^{*\top} A^{*\top} A'\right)^\top,\\ &= A^{*\top} A'X' + A'^\top A^* X^*,\\ &= 2A^\top AX.\end{aligned}$$

Accordingly, the complete derivative of the cost function becomes

$$\begin{aligned}\frac{\mathrm{d}\mathcal{L}}{\mathrm{d}X} &= 0 - \left(B^\top A\right)^\top - A^\top B + 2A^\top AX,\\ &= -2A^\top B + 2A^\top AX.\end{aligned}$$

Setting this derivative to zero, the solution for X becomes (3.20). □

LS is the maximum-likelihood estimator for (3.21) if the conditions which were discussed in Section 3.4 and [86, p. 254], hold. Mainly, the output corrections may not be correlated and have $\mathbb{E}(\Delta B) = 0$.

3.5.2 Weighted least squares

We have already discussed WLS in Section 3.4.2, where iteratively WLS was used within M-estimators. The cost function of WLS offers more flexibility than LS due to the introduced scaling matrix and becomes

$$\min_X \left\| \sqrt{W}(B - AX) \right\|_2^2, \tag{3.22}$$

where W is a positive diagonal matrix of dimension $W \in \mathbb{R}^{m \times m}$. The WLS closed-form solution becomes

$$\widehat{X} = (A^\top WA)^{-1} A^\top WB \tag{3.23}$$

[86, p. 279] and we can see that LS is a special case of WLS when the scaling matrix is equal to the identity matrix ($W = I$).

3.5.3 Recursive least squares

The RLS algorithm in Algorithm 3.1 is the recursive version of WLS with a special form of scaling matrix. The diagonal elements of the scaling matrix decrease exponentionally over time, which is also known as exponentionally forgetting. Hence, the scaling matrix has the form

$$\operatorname{diag}(W) = \begin{bmatrix} \lambda^{t-1} & \dots & \lambda^2 & \lambda^1 & \lambda^0 \end{bmatrix}^\top$$

and the RLS cost function becomes

$$\min_X \sum_{i=1}^{m} \lambda^{m-i} (B_i - A_i X)^2 \tag{3.24}$$

[100, pp. 363, 364]. The forgetting factor (λ) ($0 \ll \lambda \leq 1$) controls how much weight is given to old data. If we adjust $\lambda = 1$, RLS solves the LS cost function (3.9), where all data is considered with equal weight. Figure 3.3 showed already how λ evolves over time if $\lambda < 1$.

Algorithm 3.1: Recursive least squares (RLS)

for $t \leftarrow 1$ to m do
  input: $\widehat{X}_{t-1}, P_{t-1}, A_t, B_t, \lambda$
  $L_t = (P_{t-1}A_t^\top)\left(\lambda + A_t P_{t-1} A_t^\top\right)^{-1}$
  $\widehat{X}_t = \widehat{X}_{t-1} + L_t\left(B_t - A_t\widehat{X}_{t-1}\right)$
  $P_t = (I - L_t A_t)P_{t-1}\frac{1}{\lambda}$
  output: $\widehat{X}_t, P_t$

Ljung [100, pp. 363–365] gives the transition from (3.22) into Algorithm 3.1 by using the matrix inversion lemma, whereas Isermann and Münchhof [86, pp. 281–283] derive Algorithm 3.1 by iterative inserting new data and updating the WLS solution in (3.23).

If RLS is just another algorithm which does the same as LS or WLS with a special design of W, why do we discuss RLS and various more recursive estimators in the following? The reasons why we should favor *recursive* estimators are:

1. the number of mathematical operations does not depend on m, hence we can process infinite large data sets;
2. the required storage is fixed and known from $t = 0\,\mathrm{s}$ onwards;
3. we get a solution $\widehat{X}$ for each time-step t;
4. recursive algorithms can follow and track *time-varying* systems;
5. and because of that, recursive algorithms are ideal for real-time applications, such as vehicle control.

Recursive estimators can be shown as individual equations or as algorithms. Rather than individual equations, I favor to present recursive estimators as algorithms because recursive methods require initialization of some values and constants, a certain sequence of computations, and deliver results at each time-step.

Algorithm 3.2: Recursive M-estimator (RLM)

for $t \leftarrow 1$ to m do
  input: $\widehat{X}_{t-1}, P_{t-1}, A_t, B_t, W_{t,t}, \lambda$
  $L_t = \left(W_{t,t} P_{t-1} A_t^\top\right)\left(\lambda + W_{t,t} A_t P_{t-1} A_t^\top\right)^{-1}$
  $\widehat{X}_t = \widehat{X}_{t-1} + L_t\left(B_t - A_t \widehat{X}_{t-1}\right)$
  $P_t = (I - L_t A_t) P_{t-1} \frac{1}{\lambda}$
  output: $\widehat{X}_t, P_t$

For instance, we can see in Algorithm 3.1 that RLS requires the initial values $\widehat{X}_{t-1}, P_{t-1}$ (both appear at the input line); a constant λ (no index t) and provides $\widehat{X}_t, P_t$ as result for each time-step.

3.5.4 Recursive M-estimator

The end of Section 3.4.2 indicated already that M-estimators, WLS, and RLS are connected and we have seen the need for robust estimators in Section 3.4. The recursive M-estimator (RLM) was discovered more than once. Dai and Sinha [40] is, to the best of my knowledge, the first reference that gives an RLM algorithm without forgetting factor. They were followed by Zou, Chan, and Ng [199], who proposed the same core algorithm but included a forgetting factor in the RLM cost function

$$\min_X \sum_{i=1}^{m} \lambda^{m-i} \rho(B_i - A_i X). \tag{3.25}$$

The derivation from (3.25) into Algorithm 3.2 was done by [199] through the matrix inversion lemma. Algorithm 3.2 differs from Algorithm 3.1 in the introduced time-dependent weight $W_{t,t}$, that is a scalar, denoted by the double indexing. As discussed in Section 3.4.2, RLM covers RLS as special case if the weight is adjusted to $W_{t,t} = 1, \forall t$. It is straight forward to apply (3.15b), (3.17b) or other influence functions to compute the weight with $W_{t,t} = \psi_t / \Delta B_t$.

However, we know from (3.13), that we need a simultaneous robust method for the estimated scale. Zou, Chan, and Ng [199] suggested

$$\widehat{\sigma}_t^2 = \lambda\widehat{\sigma}_{t-1}^2 + (1-\lambda)1.483\left(1 + \frac{5}{j-n}\right)\operatorname*{med}_{i=0}^{j} \Delta B_{t-i}^2, \tag{3.26}$$

which is a sliding-window version of the first equation in [142, p. 44]. However, (3.26) requires to store j squared output corrections in a memory and to compute the median of them. This procedure is not computational efficient because of the memory size j which is needed to compute the estimated scale. Moreover, the median requires to sort all j entries. Consequently, one would always have to balance between large-j, which is good for statistics and small-j, which is computational cheaper. Hence, (3.26) leads into a dilemma. The same problem arises if we would use the robust MAD from (3.14) to compute the estimated scale.

Rousseeuw and Bassett [140] suggested to replace the median with the repeated median for large data sets. The repeated median performs nested median operations with hierarchical ordered small buffers and is computational more efficient than the conventional median.

Although the repeated median is an improvement over the conventional median, we should seek for a robust scale estimator which has comparable computational load as the non-robust standard deviation (3.27a) and its recursive version (shown as variance in (3.27b))

$$\widehat{\sigma} = \sqrt{\frac{1}{m-n}\sum_{i=1}^{m} \Delta B_i^2}, \tag{3.27a}$$

$$\widehat{\sigma}_t^2 = \lambda\widehat{\sigma}_{t-1}^2 + (1-\lambda)\Delta B_t^2, \tag{3.27b}$$

[140, p. 40] and [199], respectively. More explanation for the recursive standard deviation can be found in [35, 191].

Bylander and Rosen [29] presented an recursive median algorithm which grounds on perceptron learning-algorithm [138]. The idea is to use some estimated median from the previous time-step, compare it with incoming data, and correct the estimate in direction of steepest descent with the learn rate (η). Further, Bylander and Rosen [29] explain a slight modification which delivers estimated quantiles ($\widehat{Q}$) for any desired probability (p) (the median is the quantile on 0.5 probability) and this modification makes their algorithm general. The estimated quantile of the squared output correction is gained by

$$\widehat{Q}_t = \begin{cases} \widehat{Q}_{t-1} + 2\eta p & \text{if } \Delta B_t^2 > \widehat{Q}_{t-1}, \\ \widehat{Q}_{t-1} & \text{if } \Delta B_t^2 = \widehat{Q}_{t-1}, \\ \widehat{Q}_{t-1} - 2\eta(1-p) & \text{if } \Delta B_t^2 < \widehat{Q}_{t-1}, \end{cases} \tag{3.28}$$

and delivers the recursive median, if we adjust the probability to $p = 0.5$. The learn rate (η) ranges typically between 0.001 to 0.01 and determines the convergence rate of the estimator.

An equivalent formulation without piece-wise definition is

$$\widehat{Q}_t = \widehat{Q}_{t-1} + \eta|\operatorname{sgn}(\Delta B_t^2 - \widehat{Q}_{t-1})|(\operatorname{sgn}(\Delta B_t^2 - \widehat{Q}_{t-1}) + 2p - 1). \tag{3.29}$$

Replacing everything right of the scaling factor 1.483 in (3.26) with $\widehat{Q}_t$ of (3.29) yields the desired robust estimated scale which is required to run RLM as shown in Algorithm 3.3.

First, the normalized a priori output correction (ΔB_t) is computed with estimates of $\widehat{X}$ from the previous time-step in Line 2. Second, Line 3 and Line 4 yield the weight based on the a priori output correction and the estimated scale from the previous time step. Third, RLM gives $\widehat{X}$ and P for the current time-step. Fourth, the a posteriori output correction in Line 7 is used to update the robust estimated scale in Line 8–Line 9.

Algorithm 3.3: RLM with robust estimated scale

1 **for** $t \leftarrow 1$ **to** m **do**

  input: $\widehat{X}_{t-1}, P_{t-1}, \widehat{\sigma}^2_{t-1}, \widehat{Q}_{t-1}, A_t, B_t, \lambda, \eta, p = 0.5$

2   $\Delta B_t = (B_t - A_t\widehat{X}_{t-1})$

3   $\psi_t(\Delta B_t/\widehat{\sigma}_{t-1}) = \begin{cases} \text{(3.15b), or} \\ \text{(3.17b)} \end{cases}$

4   $W_{t,t} = \dfrac{\psi_t}{\Delta B_t/\widehat{\sigma}_{t-1}}$

5   function call

   input: $\widehat{X}_{t-1}, P_{t-1}, A_t, B_t, W_{t,t}, \lambda$

6   RLM (Algorithm 3.2)

   output: $\widehat{X}_t, P_t$

7   $\Delta B_t = B_t - A_t\widehat{X}_t$

8   $\widehat{Q}_t = \widehat{Q}_{t-1} + \eta|\text{sgn}(\Delta B_t^2 - \widehat{Q}_{t-1})|(\text{sgn}(\Delta B_t^2 - \widehat{Q}_{t-1}) + 2p - 1)$

9   $\widehat{\sigma}_t^2 = \lambda\widehat{\sigma}^2_{t-1} + (1-\lambda)1.483\widehat{Q}_t$

  output: $\widehat{X}_t, P_t, \widehat{\sigma}_t^2, \widehat{Q}_t$

Algorithm 3.3 gives the same result as RLS if we use the influence function of LS (3.16b) in Line 3. Further, we can simplify Algorithm 3.3 if a user-defined fixed estimated scale is given. Then, Line 7–Line 9 can be omitted.

Note that Algorithm 3.3 grounds on the implicit assumption that one iteration per time-step is sufficient. In other words, Algorithm 3.3 is a suboptimal estimator for (3.25). However, Algorithm 3.3 can be modified into an optimal estimator through multiple iterations of Line 2–Line 9 for each t until $\widehat{X}_t$ converges. Remember that the cost function in (3.13) requires iterative procedures (iteratively WLS), as discussed in Section 3.4.2.

3.5.5 Recursive regularized M-estimator

Remember the reasons for applying regularization which were discussed in Section 3.3 and let us write once again the cost function (3.8b) for LS with Tikhonov regularization as sum with exponential forgetting

$$\min_X \sum_{i=1}^{m} \left(\lambda^{m-i}(B_i - A_i X)^2\right) + \kappa \|\mathcal{R}X\|_2^2, \tag{3.30}$$

where $\mathcal{R} = I$. Now let us replace the squares in (3.30) with the ρ-function to derive the recursive regularized M-estimator cost function

$$\min_X \sum_{i=1}^{m} \left(\lambda^{m-i}\rho(B_i - A_i X)\right) + \kappa \|\mathcal{R}X\|_2^2. \tag{3.31}$$

The regularizer (right hand side in (3.31)) adds a bias towards zero and prevents $\widehat{X}$ becoming suspicious large. Hence, the estimation variance (or uncertainty) is reduced during poor excitation [186].
Van Waterschoot, Rombouts, and Moonen [186] give the batch solution for Tikhonov regularized LS (3.8b)

$$\widehat{X} = (A^\top A + \kappa I)^{-1} A^\top B, \tag{3.32}$$

where, compared with the LS solution (3.20), a scaled identity matrix is added to the input covariance matrix (R) ($R = A^\top A$). Further, the recursions for the input covariance matrix and estimated parameters become

$$R_t = \lambda R_{t-1} + A_t^\top A_t + (1-\lambda)\kappa I, \tag{3.33a}$$

$$\widehat{X}_t = \widehat{X}_{t-1} + R_t^{-1}\left(A_t^\top (B_t - A_t \widehat{X}_{t-1}) - (1-\lambda)\kappa \widehat{X}_{t-1}\right) \tag{3.33b}$$

[186]. However, the inversion of R in (3.33b) should be avoided by the matrix inversion lemma similar to RLS and RLM in Algorithm 3.1 and Algorithm 3.2 respectively. Additionally, let us condense the term $(1-\lambda)\kappa$ into a scaled κ.

Algorithm 3.4: Recursive regularized M-estimator (RRLM)

for $t \leftarrow 1$ to m do
    input: $\widehat{X}_{t-1}, P_{t-1}, A_t, B_t, W_{t,t}, \lambda, \kappa$
    $L_t = \left(W_{t,t} P_{t-1} A_t^\top\right)\left(\lambda + W_{t,t} A_t P_{t-1} A_t^\top\right)^{-1}$
    $P_t = (I - L_t A_t) P_{t-1} \frac{1}{\lambda}$
    $P_t = P_t (I + \kappa P_t)^{-1}$
    switch *type* do
        case *Levenberg-Marquardt regularization* do
            $\widehat{X}_t = \widehat{X}_{t-1} + L_t\left(B_t - A_t \widehat{X}_{t-1}\right)$
        case *Tikhonov regularization* do
            $\widehat{X}_t = \widehat{X}_{t-1} + L_t\left(B_t - A_t \widehat{X}_{t-1}\right) - P_t \kappa \widehat{X}_{t-1}$
    output: $\widehat{X}_t, P_t$

Gunnarsson [68] showed that the inversion of the input covariance matrix $P = R^{-1}$ of (3.33a) leads to a normalization of the covariance matrix P after the P-update

$$P_t' = P_t (I + \kappa P_t)^{-1}, \tag{3.34}$$

where P_t' is the regularized covariance matrix. The combination of (3.33), (3.34), and Algorithm 3.2 leads to the recursive regularized M-estimator (RRLM) with the two regularization methods Levenberg-Marquardt regularization and Tikhonov regularization in Algorithm 3.4.

The Levenberg-Marquardt regularization (also shown in [186]) differs from Tikhonov regularization in the computation of the parameter update. Levenberg-Marquardt regularization causes that the parameters remain at their values during poor excitation, whereas Tikhonov regularization causes that the parameters tend towards zero during poor excitation. RRLM turns into RLM if $\kappa = 0$. Hence, RLM and RLS are included as special cases in RRLM and we can substitute RLM (Algorithm 3.2) in Line 6 of Algorithm 3.3 with RRLM to create a flexible recursive regularized robust estimator for the linear multi-input-single-output output-error model.

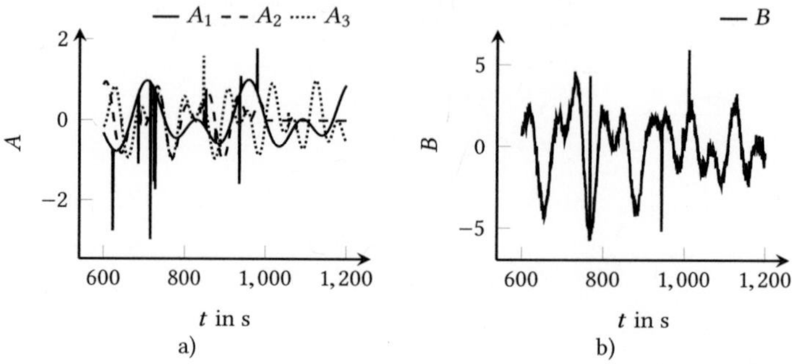

Figure 3.7: Measured inputs (A) and measured output (B) modeled as sine waves [altered from 133]. A_1, A_2, and A_3 are noise free, whereas B has additive white Gaussian noise. A_2 shows no excitation between 1000 s to 6000 s. All signals contain outliers.

3.5.6 Experiments

Let us conduct a sophisticated experiment to study the properties of RRLM in conjunction with the robust estimated scale with different settings of Algorithm 3.4 inside Algorithm 3.3. Figure 3.7 shows the measured inputs and measured output which were generated with

$$
\begin{aligned}
t &= \begin{bmatrix} 1 & 2 & 3 & \dots & m \end{bmatrix}^\top, \; m = 10\,000\,\mathrm{s}, \\
\overline{A}_{:,1} &= \sin(2\pi\, t\, 0.006) \sin(2\pi\, t\, 0.006/3.3), \\
\overline{A}_{:,2} &= \begin{cases} \sin(2\pi\, t\, 0.012) \sin(2\pi\, t\, 0.012/3.3) & 1\,\mathrm{s} \le t < 1000\,\mathrm{s} \\ 0 & 1000\,\mathrm{s} \le t < 6000\,\mathrm{s} \\ \sin(2\pi(t-5000)0.012) & \\ \qquad \sin(2\pi(t-5000)0.012/3.3) & 6000\,\mathrm{s} \le t \le 10\,000\,\mathrm{s} \end{cases}, \\
\overline{A}_{:,3} &= \sin(2\pi\, t\, 0.014) \sin(2\pi\, t\, 0.014/3.3),
\end{aligned}
$$

$$\overline{X}_{1:3,:} = \begin{cases} \begin{bmatrix} 1 & 2 & 3 \end{bmatrix}^\top & 1\,\mathrm{s} \leq t < 5000\,\mathrm{s} \\ \begin{bmatrix} 1.5 & 2 & 3 \end{bmatrix}^\top & 5000\,\mathrm{s} < t \leq 10\,000\,\mathrm{s} \end{cases},$$

$$\overline{B} = \overline{A} \odot \overline{X}$$

[altered from 133]. All measured inputs are noise free ($A = \overline{A}$), whereas B contains additive white Gaussian noise with zero mean and 0.1 variance. Further, A and B contain 5 % outliers, that were built from uniform random numbers ranging from −4 to 4. A_2 shows no excitation (remains constant) between 1000 s to 6000 s and $\odot$ is the element-wise product which is also known as Hadamard product. The step change in $\overline{X}_1$ at $t = 5000$ s creates a time-varying system.

Experiment 3.3

Call the function linearMISOoe() four times with the strings 'RLS', 'RLM', 'Levenberg-Marquardt-RRLM' and 'Tikhonov-RRLM' to compute the four parameter estimates in Figure 3.8. Algorithm 3.3 with Algorithm 3.4 in Line 6 (RRLM with robust estimated scale) will be executed with $\lambda = 0.995$ for all cases. The regularization parameter is adjusted to $\kappa = 0$ for 'RLS' and 'RLM', $\kappa = 0.1$ for 'Levenberg-Marquardt-RRLM' and $\kappa = 0.001$ for 'Tikhonov-RRLM'. The influence function is (3.16b) for 'RLS' and (3.15b) for all other cases.

Figure 3.8a shows RRLM in a setting that is identical to basic RLS. The solution is heavily distorted through outliers in A and B. Hence, RLS is not able to produce meaningful results for Experiment 3.3.

The result of the robust RLM estimator with Huber's cost function is shown in Figure 3.8b. The estimates are more stable than in Figure 3.8a. However, $\widehat{X}_2$ becomes uncertain for $t = 5000$ s to 6000 s. This uncertainty is known as wind-up problem. A_2 was set to zero for $t = 1000$ s to 6000 s, which leads to an ill-posed problem. However, the estimator keeps an exponentially decreasing amount of old information until $t \approx 5000$ s, which explains the delay between the beginning of zero excitation of A_2 at $t = 1000$ s and the wind-up of $\widehat{X}_2$ at $t \geq 5000$ s. Compare also with Figure 3.3.

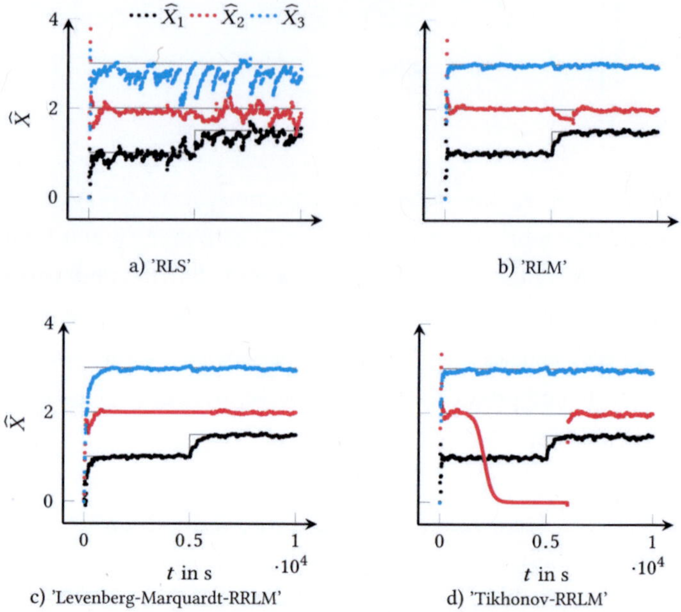

Figure 3.8: Estimated parameters for four different settings of RRLM. Nominal values are shown with — lines.

Figure 3.8c and Figure 3.8d show the result of RRLM with two different regularizations. The difference between Levenberg-Marquardt regularization in Figure 3.8c and Tikhonov regularization in Figure 3.8d is that Tikhonov regularization forces $\widehat{X}_2$ towards zero when $A_2 \approx 0$ between 1000 s to 6000 s, whereas Levenberg-Marquardt regularization keeps $\widehat{X}_2$ in this period constant. Tikhonov regularization is in this case a kind of *feature selection*, because any $\widehat{X}_2$ would lead to $A_2\widehat{X}_2 \approx 0$ if $A_2 \approx 0$. Hence, we could discard A_2 and $\widehat{X}_2$ for $t = 1000$ s to 6000 s without loosing accuracy of the model. The same conclusions can be derived from Figure 3.9 where the SEVN is shown for all four settings of Experiment 3.3.

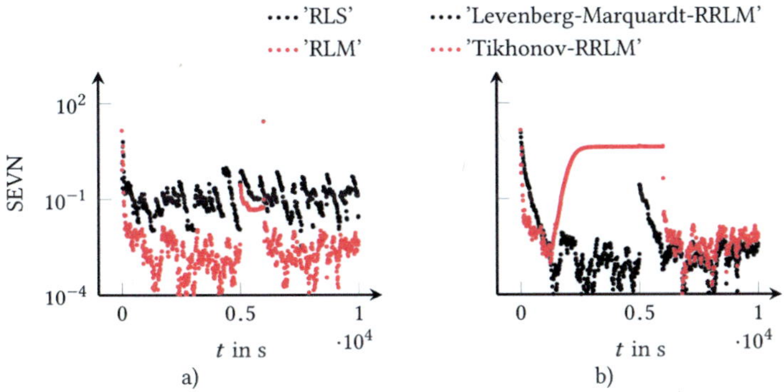

Figure 3.9: Squared error vector norm for four different settings of RRLM.

However, the SEVN (3.2) allows to compare various estimators without the need to show all parameters explicitly. Hence, SEVN will be used as performance index for the parameter error from this point onwards.

Figure 3.9a shows that the squared error vector norm of RLM is significantly lower than RLS. Furthermore, the wind-up around $t = 5000$ s is well visible. The two different regularization methods of RRLM in Figure 3.9b lead to a more stable solution when A_2 has no excitation. However, in terms of SEVN, the Levenberg-Marquardt regularization is the far better choice in comparison to Tikhonov regularization.

Finally, Figure 3.10 gives the estimated scale in Figure 3.10a as well as the weight in Figure 3.10b for the RRLM estimator with Levenberg-Marquardt regularization. The robust estimated scale of Line 8–Line 9 in Algorithm 3.3 yields sufficient accurate scale estimates with fast convergence, see Figure 3.10a.

To sum up, we can conclude that M-estimators in conjunction with an appropriate regularization method lead to superior accurate parameter estimates, although M-estimators are not robust against bad leverage points as discussed at the end of Section 3.4.2. The results of Experiment 3.3 indicate that M-estimators are sufficient robust for practical needs.

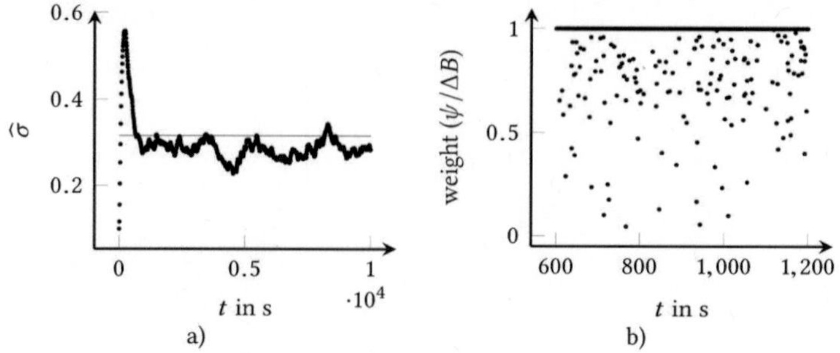

Figure 3.10: Estimated scale in Figure 3.10a and weight in Figure 3.10b of the RRLM estimator with the setting 'Levenberg-Marquardt-RRLM'.

3.6 Linear multi-input-single-output errors-in-variables model

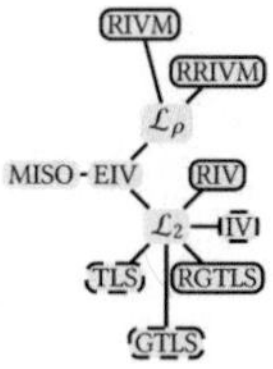

Let us switch now to the next branch of Figure 3.1, as denoted by the figure on the margin. The linear multi-input-single-output errors-invariables (EIV) model

$$AX \approx B, \quad A = \overline{A} + \widetilde{A}, \quad B = \overline{B} + \widetilde{B}, \tag{3.35a}$$

$$B = (A - \Delta A)\widehat{X} + \Delta B, \tag{3.35b}$$

considers in contrast to (3.18) input noise ($\widetilde{A}$) and output noise ($\widetilde{B}$). In other words, it is assumed that all signals have some uncertainty. Therefore, problem (3.35) is an unconstrained perturbation problem, more general than (3.18), and comprises the latter one as special case. Generally speaking, errors-in-variables estimators are more complex than output-error estimators. This drawback is compensated by the reduced parameter bias of errors-in-variables estimators that has been observed in numerous studies if the constrained perturbation assumption of (3.18) does not hold [37, 41, 49, 53, 94, 106, 116, 133, 134, 164, 166], [73, p. 116], [86, pp. 302–304], and [185, p. 5].

________ *Reused material:* Errors-in-variables estimators can be divided into two classes. The first class requires knowledge of the noise covariance matrix. The bias-compensating RLS algorithm by [44] adds a correction term which is built from the noise variance and a correlation matrix and adjusts the estimates. Furthermore, numerous recursive TLS (RTLS) algorithms based on the minimization of the Rayleigh quotient were proposed by [7, 41, 54, 55, 99]. All of these methods try to solve the errors-in-variables problem [(3.35)] with a cost function that considers data corrections in all elements of the augmented data (Z).
In the second class, the bias in the estimated parameters is reduced by properly chosen instruments ($\mathfrak{A}$). A recursive total instrumental variables (RTIV) estimator was introduced by [54]. The main advantage of instrumental variables (IV) methods is that no knowledge of the noise covariance matrix is required [This reused material has been reformatted for uniformity. ©2014 IFAC. 133]. ________
However, IV methods impose other assumptions. Essentially, the instruments should be chosen in a way that they are uncorrelated with input noise and output noise, but maximally correlated with the measured inputs. The simplest way to yield proper instruments is to use time-delayed measured inputs. Anyhow, we will discuss IV methods in detail in Section 3.6.4.

3.6.1 Total least squares (TLS)*

________ *Reused material:* Markovsky and Van Huffel [106] pointed out that TLS searches for optimal data corrections $\begin{bmatrix}\Delta A & \Delta B\end{bmatrix} := \begin{bmatrix}A & B\end{bmatrix} - \begin{bmatrix}\widehat{A} & \widehat{B}\end{bmatrix}$ (3.36), where $\|\cdot\|_{\mathrm{F}}$ denotes the Frobenius norm.

$$\min_{X \in \mathbb{R}^{q \times d}, [\widehat{A}\ \widehat{B}] \in \mathbb{R}^{m \times q}} \left\| \begin{bmatrix}A & B\end{bmatrix} - \begin{bmatrix}\widehat{A} & \widehat{B}\end{bmatrix} \right\|_{\mathrm{F}} \quad \text{s.t.} \quad \begin{bmatrix}\widehat{A} & \widehat{B}\end{bmatrix} \begin{bmatrix} X \\ -I \end{bmatrix} = 0 \qquad (3.36)$$

The approximate solution of the overdetermined system of equations is $\widehat{A}\widehat{X} = \widehat{B}$. If the noise is independently identically distributed with zero mean and a covariance matrix

$$\widetilde{P} = \operatorname{cov}\left(\begin{bmatrix} \widetilde{A} & \widetilde{B} \end{bmatrix}\right) = \sigma^2 I, \tag{3.37}$$

equal to the identity matrix up to σ^2, TLS is the maximum-likelihood estimator for (3.35) [106]. Note that σ^2 is an unknown scalar which does not affect the TLS correction. [...]

The solution of the basic TLS requires the svd$(\cdot)$ (3.38) of the augmented data $Z = \begin{bmatrix} A & B \end{bmatrix}$, where $Z \in \mathbb{R}^{m \times q}$. The matrices $U \in \mathbb{R}^{m \times m}$ and $V \in \mathbb{R}^{q \times q}$ are orthonormal unitary $U^\top U = I$, $V^\top V = I$ and their columns are called the left and right singular vectors, respectively. The non-negative diagonal matrix $S \in \mathbb{R}^{m \times q}$ contains the singular values of Z in decreasing order.

$$Z = USV^\top, \quad U^\top Z V = S, \quad S = \operatorname{diag}(S_{1,1}, \ldots, S_{q,q}) \tag{3.38}$$

Algorithm 3.5 provides the required computations for the basic TLS solution. First, compute the svd of Z (Algorithm 3.5 Line 3). After that, partition V (Algorithm 3.5 Line 4), and finally, compute the parameter estimate $\widehat{X}$ according to Algorithm 3.5 Line 5 [185, p. 37]. Note that only V is needed from the svd in Algorithm 3.5 Line 3 to compute the parameter estimate $\widehat{X}$ in Algorithm 3.5 Line 5. The solution is generic if V_{22} is non-singular. In our case with $d = 1$, it is generic if $V_{22} \neq 0$. Furthermore, the solution is unique if $S_{n,n} \neq S_{q,q}$ [106]. Extensions to the non-generic and non-unique case are categorized in [185, p. 50].

In our opinion, the covariance information $\operatorname{cov}\left(\widehat{X}\right)$ of the estimate $\widehat{X}$ is as important as the estimate itself. This estimation is a challenging task in TLS and is discussed only insufficiently in the TLS literature. Van Huffel and Vandewalle [185, p. 242] provide an approximate covariance formula which we integrate in Algorithm 3.5 Line 7.

The augmented correction (ΔZ) is given with

$$\Delta Z = \begin{bmatrix} \Delta A & \Delta B \end{bmatrix} = S_{q,q} U_{:,q} V_{:,q}^{\top}, \tag{3.39}$$

[185, p. 35]. We found that

$$\begin{bmatrix} \Delta A & \Delta B \end{bmatrix} \approx \begin{bmatrix} A & B \end{bmatrix} \begin{bmatrix} V_{12} \\ V_{22} \end{bmatrix} \begin{bmatrix} V_{12}^{\top} & V_{22}^{\top} \end{bmatrix} \tag{3.40}$$

is a more convenient form which has derived in accordance with [183, p. 435]. During our simulations, the error between the exact form (3.39) and the approximate (3.40) was in the range of machine precision. Note that (3.40) has the advantage that the costly matrix U is not required. Finally, the approximate data is $\begin{bmatrix} \widehat{A} & \widehat{B} \end{bmatrix} = \begin{bmatrix} A & B \end{bmatrix} - \begin{bmatrix} \Delta A & \Delta B \end{bmatrix}$. [...]
Figure 3.11 visualizes the difference between LS and TLS. While LS corrects the data vertically and assumes that A is exactly known, TLS performs perpendicular data corrections. That is also the reason why TLS is sometimes called orthogonal regression [This reused material has been reformatted for uniformity. ©2013 IEEE. 135, pp. 269–270]. ________

3.6.2 Generalized total least squares (GTLS)*

________ *Reused material:* So far, TLS seems to be the superior method, due to the more realistic unconstrained perturbation model (3.35). However, TLS requires quite restrictive conditions for maximum-likelihood characteristics. In practice, it is unlikely that all errors are uncorrelated and equally sized as required by (3.37).
Generalizations of basic TLS can deal with column-wise or row-wise correlated noise and unequally sized error covariance matrices [184]. Markovsky et al. [107] introduced an element-wise weighted TLS method and accepted the drawback of losing a closed-form solution.

Algorithm 3.5: Total least squares (TLS)

input: A, B

batch

$Z = \begin{bmatrix} A & B \end{bmatrix}$

$USV^\top = \text{svd}(Z)$

$$V := \begin{matrix} & n & d \\ n & V_{11} & V_{12} \\ d & V_{21} & V_{22} \end{matrix}$$

$\widehat{X} = -V_{12}V_{22}{}^{-1}$

$\widehat{\sigma}^2 \approx \dfrac{S_{q,q}}{m}$

$\text{cov}(\widehat{X}) \approx \left(1 + \left\|\widehat{X}\right\|_2^2\right)\widehat{\sigma}^2(A^\top A - m\widehat{\sigma}^2 I)^{-1}$

output: $\widehat{X}$, $\text{cov}\left(\widehat{X}\right)$

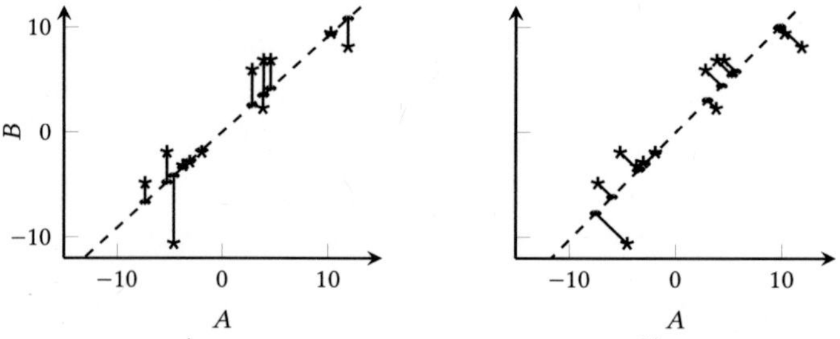

a) b)

Figure 3.11: Data fitting with LS in Figure 3.11a and TLS in Figure 3.11b. $\star$ shows the data $\begin{bmatrix} A & B \end{bmatrix}$, ∙ shows approximations $\begin{bmatrix} \widehat{A} & \widehat{B} \end{bmatrix}$, - - - shows the estimated model and —— shows the corrections $\begin{bmatrix} \Delta A & \Delta B \end{bmatrix}$. [This reused material has been reformatted for uniformity. ©2013 IEEE. 135].

Apart from svd-based TLS methods, Schaffrin and Wieser [153] introduced an element-wise weighted TLS method based non-linear Lagrange functions and Shen, Li, and Chen [159] solved this problem with a Newton-Gauss-based scheme.

Schuermans et al. [156] provide the simplest kind of generalized total least squares (GTLS) scheme through rescaling the data in a way that the noise covariance matrix meets the form required by TLS (3.37). This data scaling is performed from Line 2–Line 4 in Algorithm 3.6. The Cholesky factorization (chol(·)) of the right scaling matrix (W_r) in Line 2 is used to transform the data into a new space in Line 4. The basic TLS algorithm is used as nested function in Algorithm 3.6 Line 6 to compute the parameter estimates $\widehat{X}'$ in the transformed space. Finally, Algorithm 3.6 Line 7 converts $\widehat{X}'$ back in the original space [This reused material has been reformatted for uniformity. ©2013 IEEE. 135, pp. 270–271]. ________

The conversion of $\operatorname{cov}\left(\widehat{X}'\right)$ into $\operatorname{cov}\left(\widehat{X}\right)$ in Algorithm 3.6 Line 8 was derived in [136] and is shown herein additionally to [156] and [135]. The cost function which corresponds to the GTLS algorithm becomes

$$\min_{X \in \mathbb{R}^{q \times d}, \widehat{Z} \in \mathbb{R}^{m \times q}} \left\| \sqrt{W_l}(Z - \widehat{Z}) W_r \right\|_F \quad \text{s.t. } \widehat{Z} \begin{bmatrix} X \\ -I \end{bmatrix} = 0,$$

where $W_l \in \mathbb{R}^{m \times m}$ is a diagonal left scaling matrix which allows row-wise data weighting. Hence, W_l is the link to exponentially forgetting as discussed in Section 3.5.3 ($\operatorname{diag}(W_l) = [\lambda^{m-1}, \ldots, \lambda^2, \lambda^1, \lambda^0]^\top$).

________ *Reused material:* Algorithm 3.6 can treat three different TLS problems:

1. If $\widetilde{P} = I$, Algorithm 3.6 works like Algorithm 3.5 in the TLS sense (errors are equally sized and uncorrelated);
2. If $\widetilde{P} = \operatorname{cov}\left(\begin{bmatrix} \widetilde{A} & \widetilde{B} \end{bmatrix}\right)$, Algorithm 3.6 acts as GTLS (errors are unequally sized and correlated);
3. If $\widetilde{P}$ is a diagonal matrix $\widetilde{P} = I \odot \operatorname{cov}\begin{bmatrix} \widetilde{A} & \widetilde{B} \end{bmatrix}$, Algorithm 3.6 computes the scaled TLS solution (errors are unequally sized and uncorrelated)

[This reused material has been reformatted for uniformity. ©2013 IEEE. 135, pp. 270–271]. ________

Algorithm 3.6: Generalized total least squares (GTLS)

input: $A, B, \widetilde{P}, W_l$

batch

$C = \text{chol}(\widetilde{P})$

$$W_r = C^{-1} := \begin{matrix} & n & d \\ n & W_{r11} & W_{r12} \\ d & 0 & W_{r22} \end{matrix}$$

$\begin{bmatrix} A' & B' \end{bmatrix} = \sqrt{W_l} \begin{bmatrix} A & B \end{bmatrix} W_r$

function call

input: A', B'

Algorithm 3.5

output: $\widehat{X}', \text{cov}(\widehat{X})$

$\widehat{X} = (W_{r11}\widehat{X}' - W_{r12})W_{r22}^{-1}$

$$\text{cov}\left(\widehat{X}\right) = \frac{W_{r11}^{\top} \text{cov}\left(\widehat{X}'\right) W_{r11}}{W_{r22}^{2}}$$

output: $\widehat{X}, \text{cov}\left(\widehat{X}'\right)$

3.6.3 Recursive GTLS

———— *Reused material:* Recursive versions of GTLS with data scaling were shown in [94, 135]. These algorithms replace the batch svd(·) in Algorithm 3.5 Line 3 with efficient svd update schemes [22, 23, 67]. A drawback of data scaling is that we cannot assume one or more measured inputs as noise-free, because a scaling with zero would neglect these measured inputs. However, these algorithms provide a closed-form solution.

The majority of algorithms use power methods, such as inverse iteration or Rayleigh quotient iteration. Davila [41] showed that the minimization of the generalized Rayleigh quotient

$$\min_{V_{:,q}} \frac{V_{:,q}^{\top}(Z^{\top}Z)V_{:,q}}{V_{:,q}^{\top}\widetilde{P}V_{:,q}} \tag{3.41}$$

provides the eigenvector $V_{:,q}$ which corresponds to the smallest eigenvalue $S_{q,q}$. And this eigenvector is involved in the GTLS solution.

Lemma 3.2. *The minimization of (3.41) results in asymptotically unbiased and consistent GTLS solution X in case of additive, zero mean, Gaussian noise and known noise covariance matrix up to a multiplier by the substitution of $V_{:,q}$ with $\begin{bmatrix} X^\top & -1 \end{bmatrix}^\top$ and the substitution of $Z^\top Z$ with the expectation operator ($\mathbb{E}$) of the sample input covariance matrix for large enough t, where*

$$R_t = \frac{1}{t} \sum_{i}^{t} Z_i^\top Z_i .$$

Proof. See Proof of Theorem 1 in [41] for the substitution of $Z^\top Z$ with $\mathbb{E}(R_t)$ and Sec. B in [55] for the substitution of $V_{:,q}$ with $\begin{bmatrix} X^\top & -1 \end{bmatrix}^\top$. □

Accordingly, the eigenvector $V_{:,q}$ can be replaced with $\begin{bmatrix} X^\top & -1 \end{bmatrix}^\top$ in (3.41) and the minimization simplifies to the constrained generalized Rayleigh quotient

$$\min_{X} \frac{\begin{bmatrix} X^\top & -1 \end{bmatrix} (Z^\top Z) \begin{bmatrix} X^\top & -1 \end{bmatrix}^\top}{\begin{bmatrix} X^\top & -1 \end{bmatrix} \widetilde{P} \begin{bmatrix} X^\top & -1 \end{bmatrix}^\top} \tag{3.42}$$

[55]. Following [63, p. 465], we can solve (3.41) with generalized inverse iteration as shown in Algorithm 3.7. The while loop in Algorithm 3.7 Line 2 does not allow a closed-form solution, but generalized inverse iteration converges in the most cases fast within a few iterations. Because of that, generalized inverse iteration is suitable for recursive algorithms assuming that for each time step one iteration is sufficient to follow the smallest eigenvector.

Algorithm 3.7: Generalized inverse iteration

input: $Z, \widetilde{P}$

$V_{:,q;t} = \begin{bmatrix}1 & 1 & \dots & 1\end{bmatrix}^\top, \qquad V_{:,q;t-1} = \begin{bmatrix}0 & 0 & \dots & 0\end{bmatrix}^\top$

while $\left\|V_{:,q;t-1} - V_{:,q;t}\right\|_2 >$ threshold do

  $V_{:,q;t-1} = V_{:,q;t}$

  $V_{:,q;t} = (Z^\top Z)^{-1}(\widetilde{P} V_{:,q;t})$

  $V_{:,q;t} = V_{:,q;t} / \left\|V_{:,q;t}\right\|_2$

$\widehat{X} = -V_{1:n,q;t} / V_{q,q;t}$

output: $\widehat{X}$

The herein proposed recursive GTLS (RGTLS) algorithm that is shown in Algorithm 3.8, is based on the optimization procedure (3.42) and the recursive update of the augmented data input covariance matrix. Apart from using Z_t instead of A_t, the update in Algorithm 3.8 Line 3 conforms with Algorithm 3.1 Line 4. The constrained generalized inverse iteration is performed in Algorithm 3.8 from Line 4–Line 5 and $\widetilde{P}$ is replaced with an estimated noise covariance matrix ($\breve{P}$). $\breve{P}$ can also be used as fixed user input in the form of (3.19) for a RLS solution or (3.37) for a RTLS solution [This reused material has been reformatted for uniformity. ©2014 IFAC. 133] ________

The following paragraphs outline a noise covariance estimator (NCE) that was introduced in [133] and grounds on the polynomial Kalman smoother (PKS) that will be discussed in Section 3.9.1.

Figure 3.12 gives the block diagram of RGTLS with NCE. The PKS are used to compute the estimated input noise ($\breve{A}$) and estimated output noise ($\breve{B}$), respectively. These estimates are used in the NCE to provide $\breve{P}_t$ as input for RGTLS. Note that RGTLS uses raw data A_t and B_t. Hence, there is no delay in the estimated parameters due to the PKS smoothing.

Algorithm 3.8: Recursive GTLS (RGTLS)

for $t \leftarrow 1$ to m do
    input: $\widehat{X}_{t-1}, P_{t-1}, Z_t, \breve{P}_t, \lambda$
    $L_t = \left(P_{t-1}Z_t^\top\right)\left(\lambda + Z_t P_{t-1} Z_t^\top\right)^{-1}$
    $P_t = (I - L_t Z_t) P_{t-1} \frac{1}{\lambda}$
    $V_{:,q;t-1} = [\widehat{X}_{t-1}^\top, -1]^\top$
    $V'_{:,q;t} = P_t(\breve{P}_t V_{:,q;t-1})$
    $\widehat{X}_t = -V'_{1:n,q;t} / V'_{q,q;t}$
    output: $\widehat{X}_t, P_t$

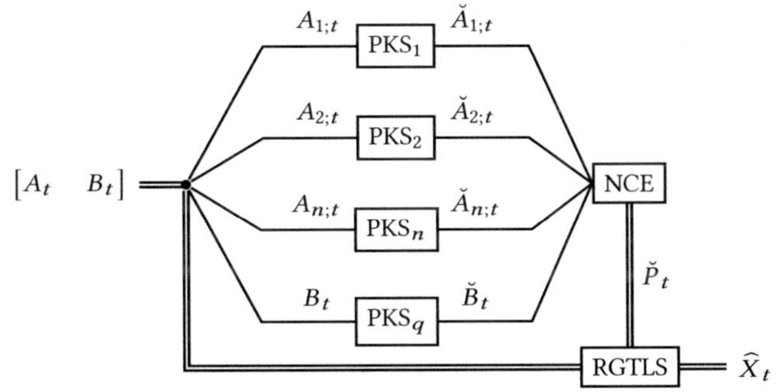

Figure 3.12: Block diagram of RGTLS with NCE. [This reused material has been reformatted for uniformity. ©2014 IFAC. 133].

Reused material: In multi-input-single-output identification, $\breve{P}$ is a square matrix with $\breve{P} \in \mathbb{R}^{q \times q}$. A simple noise covariance update formula with forgetting is

$$\breve{Z}_t = \begin{bmatrix} \breve{A}_{1;t} & \dots & \breve{A}_{n;t} & \breve{B}_t \end{bmatrix}, \tag{3.43a}$$

$$\breve{P}_t = \lambda \breve{P}_{t-1} + (1 - \lambda)\left(\breve{Z}_t^\top \breve{Z}_t\right), \tag{3.43b}$$

where (3.43b) is the multidimensional version of the noise variance estimator in [199].

Figure 3.12 shows that we need q-independent PKS to compute $\breve{Z}_t$ in (3.43a). The noisy measurement B_t in Algorithm 3.21 is $A_{1;t}$ for PKS$_1$, ..., $A_{n;t}$ for PKS$_n$ and B_t for PKS$_q$ [This reused material has been reformatted for uniformity. ©2014 IFAC. 133]. ________

Although ✎ RGTLS is a general estimator of the multi-input-single-output errors-in-variables model (3.35), two specific problems need more investigation to apply RGTLS in practice. First, a robust version of RGTLS should be derived which is similar to RLM. If we compare the RLM cost function (3.25) with the generalized Rayleigh quotient (3.41) (which is in fact the RGTLS cost function) this derivation could be straightforward.

It is more challenging to introduce adequate regularization in RGTLS. Several resources discuss that "total least squares is a deregularizing procedure" [106, p. 2288] and [62, p. 889]. As RGTLS grounds on TLS, RGTLS performs deregularization as well, which is unsuitable during poor excitation. This deregulaization can be observed if we compare the TLS solution

$$\widehat{X} = \left(A^\top A - S_{q,q}^2 I\right)^{-1} A^\top B \tag{3.44}$$

(which is equal to the shown TLS solution $\widehat{X} = -V_{12}V_{22}{}^{-1}$ in Algorithm 3.5 [185, p. 36]) with the regularized LS solution (3.32). The latter adds the term κI to $A^\top A$ whereas $S_{q,q}^2 I$ is subtracted from $A^\top A$ in (3.44).

Numerous studies indicate the significance to derive regularized TLS-based estimators [see the references 31–35 in 106] and [95, 103, 105, 160]. However, all of these references deal with batch estimators. Regularized recursive TLS-based estimators are rare. One reference in this matters is [99], where a regularized RTLS estimator was used to train neural networks.

3.6.4 Instrumental variables (IV)

Instrumental variables is a well understood and widely used method to solve the linear multi-input-single-output errors-in-variables model (3.35) [164, 165], [86, pp. 302–304], [168, pp. 260–264], [100, pp. 224–226] and [73, pp. 116-118]. As said in Section 3.6, instrumental variables belongs to the second class of errors-in-variables estimators which does not require knowledge of the noise covariance matrix. This simplification and the strong connection to basic LS (and hence to RLS, RLM and RRLM in Section 3.5.3–Section 3.5.5) are the two main advantages of instrumental variables over the more sophisticated TLS-based estimators which we discussed in Section 3.6.1–Section 3.6.3.
Isermann and Münchhof [86, p. 302] explain the instrumental variables idea as follows: The LS output correction

$$\Delta B = B - A\widehat{X}$$

is augmented with instruments ($\mathfrak{A}$), $\mathfrak{A} \in \mathbb{R}^{m \times n}$ on both sides

$$\mathfrak{A}^\top \Delta B = \mathfrak{A}^\top B - \mathfrak{A}^\top A\widehat{X},$$

where $\mathfrak{A}$ should be chosen such that $\mathfrak{A}$ is uncorrelated with the input noise and output correction [154, p. 73], $\mathbb{E}\left(\mathfrak{A}^\top \widetilde{A}\right) = 0$ and $\mathbb{E}\left(\mathfrak{A}^\top \Delta B\right) = 0$ and maximally correlated with A so that $\mathbb{E}\left(\mathfrak{A}^\top A\right)$ is positive definite. Hence,

$$0 = \mathfrak{A}^\top B - \mathfrak{A}^\top A\widehat{X},$$

which leads to the instrumental variables closed-form solution

$$\widehat{X} = \left(\mathfrak{A}^\top A\right)^{-1} \mathfrak{A}^\top B, \tag{3.45}$$

[86, p. 302], which is similar to the LS closed-form solution (3.20). Further, we can observe that (3.45) comprises LS as special case if we adjust the instruments to $\mathfrak{A} := A$.

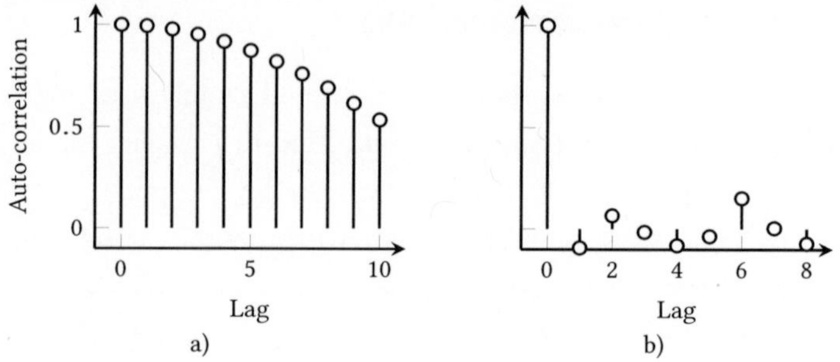

Figure 3.13: Auto-correlation plots of a sine wave in Figure 3.13a and white Gaussian noise in Figure 3.13b.

The accuracy of instrumental variables (a well reduced parameter bias compared with LS) depends strongly on proper chosen instruments and the simplest way to choose the instruments is to use delayed measured inputs or filtered inputs [167]. In some cases instrumental variables estimators might not lead to useful results or even diverge. Such cases were studied by numerical experiments in [154, pp. 185–190]. Accordingly, the instruments should fulfill the following conditions.

First, if we use delayed measured inputs as instruments, instrumental variables requires that A comes from a monotonic signal source which shows an auto-correlation plot, where the correlation coefficient decreases slowly over the time lag. Figure 3.13a shows a good suited auto-correlation of a sine wave. Instrumental variables would fail in case of A generated by Gaussian noise because $\mathbb{E}(\mathfrak{A}^\top A)$ would tend to zero which leads to the ill-posed problem. Figure 3.13b shows the auto-correlation plot of white Gaussian noise. Note the drastic drop of the correlation coefficient for lags greater then zero in Figure 3.13b.

Algorithm 3.9: Recursive IV (RIV)

1 for $t \leftarrow 1$ to m do

  input: $\widehat{X}_{t-1}, P_{t-1}, A_t, B_t, \mathfrak{A}_t, \lambda$

2 $L_t = \left(P_{t-1}\mathfrak{A}_t^\top\right)\left(\lambda + A_t P_{t-1}\mathfrak{A}_t^\top\right)^{-1}$

3 $\widehat{X}_t = \widehat{X}_{t-1} + L_t\left(B_t - A_t\widehat{X}_{t-1}\right)$

4 $P_t = (I - L_t A_t)P_{t-1}\frac{1}{\lambda}$

  output: $\widehat{X}_t, P_t$

Second, instrumental variables performs well if "the power spectrum of the noise is much wider than the power spectrum of the input" [154, p. 73]. In other words, the noise should contain approximately uniform power over all frequencies, which is characteristic for white noise. Instrumental variables perform not so well if the noise is auto-correlated which can be simulated if we pass white noise through a Butterworth filter. This second condition can be relaxed with the extended instrumental variables or overdetermined instrumental variables estimator, where the dimension of $\mathfrak{A}$ is larger than the dimension of A [57, 164], [100, p. 227] and [168, p. 262].

These two conditions depict the drawbacks of instrumental variables compared with TLS-based estimators, where the latter do not suffer under these restrictions.

3.6.5 Recursive IV

The recursive IV (RIV) algorithm is shown in Algorithm 3.9 [100, p. 369] and [168, p. 327]. The difference to RLS (Algorithm 3.1) is in Line 2, where the instruments are involved in the computation of the correction vector (L). As said, Algorithm 3.9 may be used to compute the RLS result if $\mathfrak{A}_t := A_t$. Because RIV contains RLS as special case, it is trivial to perform the same steps that led from RLS over RLM with robust estimated scale into RRLM (Section 3.5.3–Section 3.5.5) also for RIV. Because of this, the recursive IV M-estimator (RIVM) and the recursive regularized IV M-estimator (RRIVM)

Algorithm 3.10: Recursive IV M-estimator (RIVM)

for $t \leftarrow 1$ to m do
- input: $\widehat{X}_{t-1}, P_{t-1}, A_t, B_t, \mathfrak{A}_t, W_{t,t}, \lambda$
- $L_t = \left(W_{t,t} P_{t-1} \mathfrak{A}_t^{\top}\right)\left(\lambda + W_{t,t} A_t P_{t-1} \mathfrak{A}_t^{\top}\right)^{-1}$
- $\widehat{X}_t = \widehat{X}_{t-1} + L_t\left(B_t - A_t \widehat{X}_{t-1}\right)$
- $P_t = (I - L_t A_t) P_{t-1} \frac{1}{\lambda}$
- output: $\widehat{X}_t, P_t$

Algorithm 3.11: RIVM with robust estimated scale

for $t \leftarrow 1$ to m do
- input: $\widehat{X}_{t-1}, P_{t-1}, \widehat{\sigma}_{t-1}^2, \widehat{Q}_{t-1}, A_t, B_t, \mathfrak{A}_t, \lambda, \eta, p = 0.5$
- $\Delta B_t = (B_t - A_t \widehat{X}_{t-1})$
- $\psi_t(\Delta B_t / \widehat{\sigma}_{t-1}) = \begin{cases} (3.15b), \text{ or} \\ (3.17b) \end{cases}$
- $W_{t,t} = \dfrac{\psi_t}{\Delta B_t / \widehat{\sigma}_{t-1}}$
- function call
  - input: $\widehat{X}_{t-1}, P_{t-1}, A_t, B_t, \mathfrak{A}_t, W_{t,t}, \lambda$
  - RIVM (Algorithm 3.10)
  - output: $\widehat{X}_t, P_t$
- $\Delta B_t = B_t - A_t \widehat{X}_t$
- $\widehat{Q}_t = \widehat{Q}_{t-1} + \eta |\mathrm{sgn}(\Delta B_t^2 - \widehat{Q}_{t-1})| (\mathrm{sgn}(\Delta B_t^2 - \widehat{Q}_{t-1}) + 2p - 1)$
- $\widehat{\sigma}_t^2 = \lambda \widehat{\sigma}_{t-1}^2 + (1 - \lambda) 1.483 \widehat{Q}_t$
- output: $\widehat{X}_t, P_t, \widehat{\sigma}_t^2, \widehat{Q}_t$

are only given as pseudo-code with short explanation in Section 3.6.6 and Section 3.6.7, respectively.

3.6.6 Recursive IV M-estimator

The RIVM algorithm is shown in Algorithm 3.10 (compare with Algorithm 3.2) and the RIVM algorithm with robust estimated scale in Algorithm 3.11 (compare with Algorithm 3.3).

Algorithm 3.12: Recursive regularized IV M-estimator (RRIVM)

for $t \leftarrow 1$ to m do
- input: $\widehat{X}_{t-1}, P_{t-1}, A_t, B_t, \mathfrak{A}_t, W_{t,t}, \lambda, \kappa$
- $L_t = \left(W_{t,t} P_{t-1} \mathfrak{A}_t^\top\right)\left(\lambda + W_{t,t} A_t P_{t-1} \mathfrak{A}_t^\top\right)^{-1}$
- $P_t = (I - L_t A_t) P_{t-1} \frac{1}{\lambda}$
- $P_t = P_t (I + \kappa P_t)^{-1}$
- switch *type* do
  - case *Levenberg-Marquardt regularization* do
    - $\widehat{X}_t = \widehat{X}_{t-1} + L_t\left(B_t - A_t \widehat{X}_{t-1}\right)$
  - case *Tikhonov regularization* do
    - $\widehat{X}_t = \widehat{X}_{t-1} + L_t\left(B_t - A_t \widehat{X}_{t-1}\right) - P_t \kappa \widehat{X}_{t-1}$
- output: $\widehat{X}_t, P_t$

3.6.7 Recursive regularized IV M-estimator

RRIVM is shown in Algorithm 3.12 (compare with Algorithm 3.4). In accordance to Section 3.5.5, we can substitute RIVM (Algorithm 3.10) in Line 6 of Algorithm 3.11 with RRIVM to create a flexible recursive regularized robust estimator for the linear multi-input-single-output errors-in-variables model.

3.6.8 Experiments

Now we repeat the experiment of Section 3.5.6 with two different settings of RRIVM with robust estimated scale (Algorithm 3.12 inside Algorithm 3.11). All data is generated in the same way as in Section 3.5.6 apart from the input noise and the output noise. This time, also the measured inputs contain additive Gaussian noise with zero mean and the output noise has lower variance in order to meet an equivalent overall noise level compared with Section 3.5.6. The noise covariance matrix becomes a diagonal matrix (independent distributed noise) and $\widetilde{P}$ was adjusted to

$$\operatorname{diag}(\widetilde{P}) = \begin{bmatrix} 0 & 0.005 & 0.01 & 0.02 \end{bmatrix}^\top . \tag{3.46}$$

Experiment 3.4

Call the function linearMISOeiv() two times with the strings 'Levenberg-Marquardt-RRLM' and 'Levenberg-Marquardt-RRIVM' to compute the two parameter estimates in Figure 3.14. Algorithm 3.11 with Algorithm 3.12 in Line 6 (RRIVM with robust estimated scale) will be executed with $\lambda = 0.995$ for all cases and the regularization parameter is adjusted to $\kappa = 0.1$. The influence function is (3.15b). The instruments are $\mathfrak{A}_t = A_t$ for 'Levenberg-Marquardt-RRLM' and $\mathfrak{A}_t = A_{t-2}$ for 'Levenberg-Marquardt-RRIVM'. That means that the latter IV estimator used instruments with a delay of 2 s.

Estimated parameters for RRLM with Levenberg-Marquardt regularization and robust estimated scale are shown in Figure 3.14a. As we have seen from Figure 3.8, this sophisticated experiment requires robust and regularized estimators. Hence, we skip experiments with RIV and RIVM and move directly to RRIVM with robust estimated scale as second estimator in Figure 3.14b.

The main observations from Figure 3.14 are. First, $\widehat{X}_2$ diverges for 1000 s to 6000 s in Figure 3.14a. Apparently, the Levenberg-Marquardt regularization performs worse if input noise is present. Further experiments with larger regularization parameter ($\kappa = 1$) improved the results for $\widehat{X}_2$. However, the convergence rate of the other estimated parameter becomes slower in this setting. This observation is quite astonishing if we compare Figure 3.14a with Figure 3.8c, where the latter figure showed superior accuracy of 'Levenberg-Marquardt-RRLM'.

Second, we can observe biased estimates of $\widehat{X}_3$, whereas $\widehat{X}_1$ shows good accuracy in Figure 3.14a. The biased estimates can be explained by the setting of $\widetilde{P}$ in (3.46). The first entry of diag($\widetilde{P}$) is zero and matches the statistical assumption of RRLM (noise free measured inputs), whereas the third entry of diag($\widetilde{P}$) is above zero.

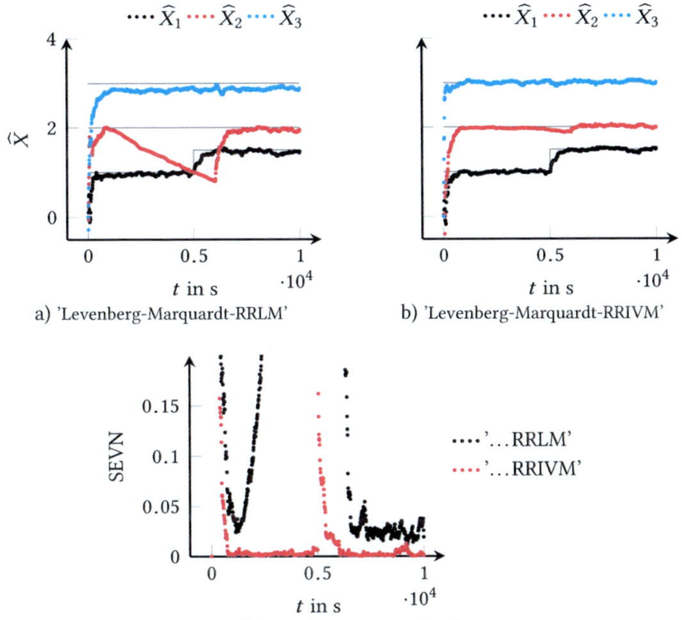

c) 'Levenberg-Marquardt-...'

Figure 3.14: Estimated parameters for Levenberg-Marquardt-RRLM and Levenberg-Marquardt-RRIVM in Figure 3.14a and Figure 3.14b, respectively. Nominal values are shown with — lines. Figure 3.14c gives the squared error vector norm.

Third, Figure 3.14b and Figure 3.14c show that the delayed instruments in 'Levenberg-Marquardt-RRIVM' ($\mathfrak{A}_t = A_{t-2}$) cause superior accuracy compared with 'Levenberg-Marquardt-RRLM'. The parameter-bias in $\widehat{X}_3$ is remarkably reduced and the efficiency of the Levenberg-Marquardt regularization for $\widehat{X}_2$ is now satisfactory.

To sum up, the introduction of instruments in RRIVM reduced the SEVN significantly for the linear multi-input-single-output errors-in-variables model.

The ✎ need for robust and regularized estimators is also the explanation why RGTLS, which was the first discussed recursive estimator for the linear multi-input-single-output errors-in-variables model in Section 3.6.3, is not shown in Figure 3.14. As said at the end of Section 3.6.3, RGTLS is currently neither robust nor regularized and cannot produce meaningful estimates for Experiment 3.4. However, if robustness and regularization is not an issue, RGTLS with NCE showed superior accuracy (smallest SEVN) compared with RLS and RTIV (the latter is a sophisticated recent recursive IV estimator by [54]) [see the experiments in 133].

> Experiment 3.5
> Download the code of [133] (http://digbib.ubka.uni-karlsruhe.de/volltexte/1000038517) and compare the squared error vector norm of RLS, RTIV and RGTLS with NCE in an experiment where robustness and regularization are not required. Note that this simpler setup is mostly to find in literature and Experiment 3.3 and Experiment 3.4 are more intricate.

Upon here, we have only considered the SEVN as performance index in all presented experiments. However, the discussion in Section 3.2.1 provided other performance indices which allow to measure the model's goodness of fit. We will now use cross-validation from Section 3.2.3 to compare the k step ahead prediction quality between 'Levenberg-Marquardt-RRLM' and 'Levenberg-Marquardt-RRIVM'.

For a fair comparison, it makes sense to split the data into training data and validation data at $t = 9000$ s because at this time, both estimators are converged. Compare Figure 3.14a and Figure 3.14b. Therefore $k = 1000$. Further, it is useful to discard the outliers in A and B. Otherwise, it would be hard to measure differences between both estimators, because outliers influence strongly all presented performance indices. In brief, the performance indices of Section 3.2.1 are not robust.

The MSE (3.3a) becomes 1.4724 for 'Levenberg-Marquardt-RRLM' and 1.5761 for 'Levenberg-Marquardt-RRIVM'. This smaller MSE of 'Levenberg-Marquardt-RRLM' is due to the used cost function of LS-related recursive

estimators. RRLM grounds on (3.25) which is a double weighted non-normalized version of MSE. All presented errors-in-variables estimators within Section 3.6 such as RRIVM exhibit cost functions which reduce the parameter-bias but not the MSE.
In conclusion, we need to decide between small SEVN (small parameter-bias) and small MSE (good prediction quality of the model) and apply one estimator of Section 3.6 or Section 3.5, respectively. This conclusion can be also found in errors-in-variables literature. For instance Van Huffel and Vandewalle [185, p. 5] mention that "[t]he errors-in-variables model is useful when the primary goal is model parameter estimation rather than prediction".

3.7 Linear random-walk output-error model

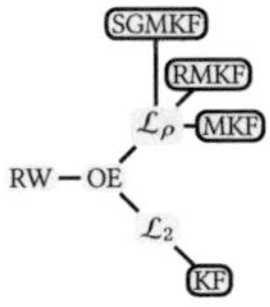

The next branch within the gray-box models of Figure 3.1 is a special kind of multi-input-single-output model that is called random-walk model. The random-walk model in [102] is designed to track parameters which vary on different rate and evolve in a stochastic manner. In other words, the state transition, which is equivalent with the temporary evolution of the states or the parameter, is not deterministic or unknown. Starting with the well known state-space model [1]

$$X_t = \mathcal{A}X_{t-1} + \mathcal{B}A_{t-1} + \Delta X_t \qquad (3.47a)$$

$$B_t = \mathcal{C}X_t + \mathcal{D}A_t + \Delta B_t, \qquad (3.47b)$$

[1] The state-space model is usually written as

$$x_{t+1} = \mathcal{A}x_t + \mathcal{B}u_t + w_t$$
$$y_t = \mathcal{C}x_t + \mathcal{D}u_t + e_t\,.$$

However, we will use the form of (3.47) to ensure a consistent nomenclature.

and setting $\mathcal{B} = \mathcal{D} = 0$, $C = A_t$ and $\mathcal{A} = I$ we get the random-walk output-error model

$$\widehat{X}_t = \widehat{X}_{t-1} + \Delta X_t \tag{3.48a}$$

$$B_t = A_t \widehat{X}_t + \Delta B_t, \tag{3.48b}$$

where ΔX_t is the parameter correction which is assumed as white Gaussian sequel with a covariance matrix $Q_t \in \mathbb{R}^{n \times n}$. The output correction (ΔB_t) is assumed as white Gaussian noise with variance R_t^1. The first power in R^1 indicates the single dimension of R^1 (multi-input-single-output model). Note that Q_t and R_t^1 can vary in time.

By setting a high value in $Q_{1,1;t}$ and lower values for all other diagonal elements in Q_t for instance, a high variation rate in X_1 is assumed, whereas the other parameters vary slower. In practice, Q_t expresses how much we believe in the estimated parameters from the previous time step and R_t^1 how much do we trust in the measured output.

3.7.1 Kalman filter

The optimal filter for solving (3.48) is the Kalman filter (KF) which is shown in Algorithm 3.13 in a form for parameter estimation [100, pp. 367–369], [168, p. 325] and [45, pp. 608–612]. However, the Kalman filter is more applied in state estimation with the full linear state-space model (3.47).

The Kalman filter is probably the most popular recursive estimator and therefore the literature on Kalman filter topics is vast. The following lemma depicts an interesting connection between the Kalman filter and recursive least squares.

Algorithm 3.13: Kalman filter (KF)

for $t \leftarrow 1$ to m do
  input: $\widehat{X}_{t-1}, P_{t-1}, A_t, B_t, R_t^1, Q_t$
  $L_t = (P_{t-1}A_t^\top)\left(R_t^1 + A_t P_{t-1} A_t^\top\right)^{-1}$
  $\widehat{X}_t = \widehat{X}_{t-1} + L_t\left(B_t - A_t\widehat{X}_{t-1}\right)$
  $P_t = (I - L_t A_t)P_{t-1} + Q_t$
  output: $\widehat{X}_t, P_t$

Lemma 3.3. *The Kalman filter is a generalization of RLS.*

Proof. If we set

$$R_t^1 = \lambda \tag{3.49a}$$

$$Q_t = P_{t-1}\left(\frac{1}{\lambda} - 1\right) + L_t A_t P_{t-1}\left(1 - \frac{1}{\lambda}\right) \tag{3.49b}$$

in Algorithm 3.13, the Kalman filter turns into RLS and Algorithm 3.13 and Algorithm 3.1 become identical. □

Hence, the Kalman filter covers RLS as special case. Moreover, it turns out that the linear random-walk output-error model in (3.48) is a generalization of the linear multi-input-single-output output-error model (3.18).

The connection between RLS and KF was extensively studied (cost functions, correspondence tables) in [151], [150, pp. 763–767], and [45, p. 614]. This similarity is important, because now findings and methods from the vast KF literature can be applied for RLS and vice versa. Hence, all evolutions of RLS which were shown from Section 3.5.3 to Section 3.5.5 for the linear multi-input-single-output output-error model, as well as the evolutions of RIV from Section 3.6.5 to Section 3.6.7 for the linear multi-input-single-output errors-in-variables model, can be applied similarly to the linear random-walk output-error model and the linear random-walk errors-in-variables model, respectively. For reasons of brevity, these similar evolutions are skipped and only the most important algorithms are given in the following.

Algorithm 3.14: M-Kalman filter (MKF)

for $t \leftarrow 1$ to m do

  input: $\widehat{X}_{t-1}, P_{t-1}, A_t, B_t, W_{t,t}, R_t^1, Q_t$

  $L_t = \left(W_{t,t} P_{t-1} A_t^\top\right)\left(R_t^1 + W_{t,t} A_t P_{t-1} A_t^\top\right)^{-1}$

  $\widehat{X}_t = \widehat{X}_{t-1} + L_t\left(B_t - A_t \widehat{X}_{t-1}\right)$

  $P_t = (I - L_t A_t) P_{t-1} + Q_t$

  output: $\widehat{X}_t, P_t$

3.7.2 M-Kalman filter

The M-Kalman filter (MKF) in Algorithm 3.14 is the robust version of the Kalman filter and can be derived similarly to the statements in Section 3.5.4. Also, MKF with robust estimated scale can be written analogously to Algorithm 3.3 if we alter the function call in Line 6 of Algorithm 3.3 from RLM into MKF and provide the required function inputs R_t^1 and Q_t instead of λ.

Durovic and Kovacevic [47] propose a robust Kalman filter which is related to MKF. However, Durovic and Kovacevic [47] provide no solution for the robust update of P. Instead, the conventional P update formula of the Kalman filter is used with the assumption that the ρ-function is almost quadratic. Hence, the robust KF in [47] is not as general as MKF in Algorithm 3.14, where a robust P update is realized and any kind of ρ-function can be applied. Moreover, the presented robust estimated scale in [47] requires a sliding window, whereas MKF with robust estimated scale does not. Remember the discussion of drawbacks for sliding windows in Section 3.5.4. Aravkin et al. [8] introduced a robust non-linear Kalman smoother and provide more references on linear and robust Kalman filters which are based on M-estimators.

Algorithm 3.15: Regularized M-Kalman filter (RMKF)

1 **for** $t \leftarrow 1$ **to** m **do**
  input: $\widehat{X}_{t-1}, P_{t-1}, A_t, B_t, W_{t,t}, R_t^1, Q_t, \kappa$
2   $L_t = \left(W_{t,t} P_{t-1} A_t^\top\right)\left(R_t^1 + W_{t,t} A_t P_{t-1} A_t^\top\right)^{-1}$
3   $P_t = (I - L_t A_t) P_{t-1} + Q_t$
4   $P_t = P_t (I + \kappa P_t)^{-1}$
5   **switch** *type* **do**
6   **case** *Levenberg-Marquardt regularization* **do**
7       $\widehat{X}_t = \widehat{X}_{t-1} + L_t\left(B_t - A_t \widehat{X}_{t-1}\right)$
8   **case** *Tikhonov regularization* **do**
9       $\widehat{X}_t = \widehat{X}_{t-1} + L_t\left(B_t - A_t \widehat{X}_{t-1}\right) - P_t \kappa \widehat{X}_{t-1}$
  output: $\widehat{X}_t, P_t$

3.7.3 Regularized M-Kalman filter

The regularized M-Kalman filter (RMKF) with Levenberg-Marquardt regularization or Tikhonov regularization can be derived similarly to RRLM (Algorithm 3.4 in Section 3.5.5) and is shown in Algorithm 3.15. The steps that lead to RMKF with robust estimated scale are now trivial. Alter the function call in Line 6 of Algorithm 3.3 from RLM into RMKF and provide the required function inputs R_t^1 and Q_t instead of λ.

3.7.4 Stenlund-Gustafsson M-Kalman filter

Another approach to deal with the wind-up problem (remember that regularization is a method to avoid the wind-up problem that is due to poor excitation) in Kalman filtering was presented in [171] and extensively studied in [50, 51].

In contrast to regularization in Section 3.7.3, where the covariance matrix (P) was normalized in Line 4 of Algorithm 3.15 with the regularization parameter κ, Stenlund and Gustafsson [171] propose to use a desired P (P_d), $P_\mathrm{d} \in \mathbb{R}^{n \times n}$ and to compute the actual Q_t depending on the direction where excitation comes. The basic idea is to regard the Kalman filter as control unit, where the goal is that the covariance matrix becomes equal with P_d, which is the desired convergence point of P [171].

Stenlund and Gustafsson [171] provide experiments which show slightly better performance of the P_d approach compared with the selective forgetting method of [124]. Selective forgetting forces the eigenvalues of the covariance matrix to lie within a given interval. More recent selective forgetting methods were proposed by [30, 31].

Although the P_d approach of [171] results in a suboptimal Kalman filter, P_d is more intuitively to use than regularization or selective forgetting. The regularized Kalman filter requires a dimensionless (and therefore rather meaningless) regularization parameter, whereas selective forgetting requires user defined minimum and maximum eigenvalues of the covariance matrix. On the other hand, P_d is the desired covariance matrix of ΔX and because of this, better interpretable as regularization or selective forgetting.

According to [171], Q_t is given by

$$Q_t = \frac{P_\mathrm{d} A_t^\top A_t P_\mathrm{d}}{R_t^1 + A_t P_\mathrm{d} A_t^\top},$$

and we can compute a robust version of Q_t with

$$Q_t = \frac{W_{t,t} P_\mathrm{d} A_t^\top A_t P_\mathrm{d}}{R_t^1 + W_{t,t} A_t P_\mathrm{d} A_t^\top}.$$

Finally, the Stenlund-Gustafsson M-Kalman filter (SGMKF) algorithm is given in Algorithm 3.16. SGMKF with robust estimated scale can be derived by modifying the function call in Line 6 of Algorithm 3.3 from RLM into SGMKF and adjust the required function inputs accordingly.

Algorithm 3.16: Stenlund-Gustafsson M-Kalman filter (SGMKF)

for $t \leftarrow 1$ to m do

  input: $\widehat{X}_{t-1}, P_{t-1}, A_t, B_t, W_{t,t}, R_t^1, P_\mathrm{d}$

  $L_t = \left(W_{t,t}P_{t-1}A_t^\top\right)\left(R_t^1 + W_{t,t}A_tP_{t-1}A_t^\top\right)^{-1}$

  $\widehat{X}_t = \widehat{X}_{t-1} + L_t\left(B_t - A_t\widehat{X}_{t-1}\right)$

  $Q_t = \dfrac{W_{t,t}P_\mathrm{d}A_t^\top A_tP_\mathrm{d}}{R_t^1 + W_{t,t}A_tP_\mathrm{d}A_t^\top}$

  $P_t = (I - L_tA_t)P_{t-1} + Q_t$

  output: $\widehat{X}_t, P_t$

3.8 Linear random-walk errors-in-variables model

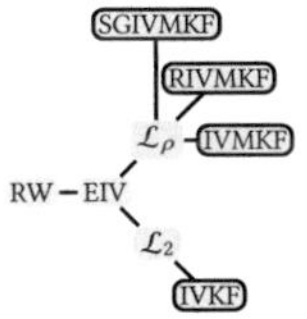

The last gray-box model of Figure 3.1 is the linear random-walk errors-in-variables model which becomes

$$\widehat{X}_t = \widehat{X}_{t-1} + \Delta X_t \tag{3.50a}$$

$$B_t = (A_t - \Delta A_t)\widehat{X}_t + \Delta B_t, \tag{3.50b}$$

and considers input corrections, whereas the random-walk output-error model (3.48) does not consider input corrections.

In accordance with Section 3.6, two classes of errors-in-variables estimators are possible to solve the random-walk errors-in-variables model. First, TLS-based estimators which require knowledge of the noise covariance matrix. Second, instrumental variables Kalman filters which do not require the noise covariance matrix.

3.8.1 Total least squares and the random-walk model*

This section grounds on collaborative and unpublished work during my research visit at the Department ELEC at Vrije Universiteit Brussel [108]. During ✎ the discussion it was found that the solution of the linear random-walk errors-in-variables model with TLS-based methods is challenging and offers research possibilities. We decided to formulate the problem first with the batch cost function

$$\min_{X,\widehat{Z}} \left({}_{W_l}\|Z - \widehat{Z}\|_{W_r} + \|X - \Lambda X\|_{W_X} \right), \tag{3.51a}$$

$$\text{s.t.} \quad \widehat{Z}_t \begin{bmatrix} X_t \\ -1 \end{bmatrix} = 0, \quad t \in \begin{bmatrix} 1 & 2 & \dots & m \end{bmatrix}. \tag{3.51b}$$

The weighting matrices $W_l \in \mathbb{R}^{m \times m}$, $W_r \in \mathbb{R}^{q \times q}$ consider the row-wise weighting with an exponential forgetting factor and the column-wise weighting of the noise covariance matrix.
Hence, $\operatorname{diag}(W_l) = \begin{bmatrix} \dots & \lambda^2 & \lambda^1 & \lambda^0 \end{bmatrix}^\top$ and $W_r = (\operatorname{chol}\widetilde{P})^{-1}$. The weighting matrix W_X considers the assumed variation in the parameters and is similar to Q_t in Section 3.7. The Λ is a short notation for a shift operator, so ΛX means a one step ahead version of X. If $X = \begin{bmatrix} X_{1;1} & X_{1;2} & \dots & X_{1;t} \end{bmatrix}$, ΛX becomes $\begin{bmatrix} X_{1;2} & X_{1;3} & \dots & X_{1;t+1} \end{bmatrix}$. We think that the recursive solution to (3.51) is a non-trivial. And so far, there is no final solution available. Possible ways to solve this problem are:

1. Solution with alternating projections
   The first idea is to solve the batch problem (3.51) with numerical optimization methods, such as alternating projections and then try to derive a online version out of that. Alternating projections requires an iterative procedure to converge to a solution. This iterative procedure is a drawback for deriving recursive algorithms. The idea is that a small number of recursive iterations is sufficient, because the optimal

solution is not so far away from the previous time step. Nevertheless, we would need to define some kind of abort threshold which is used to finish the iterations at each time step. In conclusion, a recursive algorithm based on alternating projections requires an unknown number of iterations for each time step, which is contrary to the definition of recursive or online methods.

2. Solution with robust Kalman filters
   The idea here is to use robust Kalman filters. The term robust means in this context not robust in the presence of outliers in the data, but robust in terms of model uncertainty. Sayed [149] considers the uncertain state-space model

$$
\begin{aligned}
x_t &= (\mathcal{A} + \widetilde{\mathcal{A}})x_{t-1} + (\mathcal{B} + \widetilde{\mathcal{B}})u_{t-1} + w_t,\\
y_t &= (C + \widetilde{C})x_t + e_t,
\end{aligned}
$$

   that is closely related with (3.47). Following the same arguments as in Section 3.7, we only need to consider the uncertainty in C. Due to the recursive implementation and closely related concept, this solution is superior compared with alternating projections. More references for Kalman filters which deal with model uncertainty are to find in [162, p. 313].

3.8.2 IV Kalman filter

Due to the connection between RLS and RIV on one hand (see Section 3.6.5) and RLS and KF on the other, the steps to convert the Kalman filter (Algorithm 3.13) into a IV Kalman filter (IVKF) are simple and result in the the IVKF algorithm which is shown in Algorithm 3.17.

Algorithm 3.17: IV Kalman filter (IVKF)

for $t \leftarrow 1$ to m do

  input: $\widehat{X}_{t-1}, P_{t-1}, A_t, B_t, \mathfrak{A}_t, R_t^1, Q_t$

  $L_t = \left(P_{t-1}\mathfrak{A}_t^\top\right)\left(R_t^1 + A_t P_{t-1}\mathfrak{A}_t^\top\right)^{-1}$

  $\widehat{X}_t = \widehat{X}_{t-1} + L_t\left(B_t - A_t\widehat{X}_{t-1}\right)$

  $P_t = (I - L_t A_t)P_{t-1} + Q_t$

  output: $\widehat{X}_t, P_t$

Algorithm 3.18: IV M-Kalman filter (IVMKF)

for $t \leftarrow 1$ to m do

  input: $\widehat{X}_{t-1}, P_{t-1}, A_t, B_t, \mathfrak{A}_t, W_{t,t}, R_t^1, Q_t$

  $L_t = \left(W_{t,t}P_{t-1}\mathfrak{A}_t^\top\right)\left(R_t^1 + W_{t,t}A_t P_{t-1}\mathfrak{A}_t^\top\right)^{-1}$

  $\widehat{X}_t = \widehat{X}_{t-1} + L_t\left(B_t - A_t\widehat{X}_{t-1}\right)$

  $P_t = (I - L_t A_t)P_{t-1} + Q_t$

  output: $\widehat{X}_t, P_t$

3.8.3 IV M-Kalman filter

Further, we need only small modifications to derive the robust IV M-Kalman filter (IVMKF) from MKF. Consequently, the IVMKF is presented in Algorithm 3.18 without further explanation. Also the IVMKF with robust estimated scale can be written analogously to Algorithm 3.11 if we alter the function call in Line 6 of Algorithm 3.11 from RIVM into IVMKF and provide the required function inputs R_t^1 and Q_t instead of λ.

3.8.4 Regularized IV M-Kalman filter

The regularized IV M-Kalman filter (RIVMKF) is given in Algorithm 3.19 and the RIVMKF with robust estimated scale can be written by modifying the function call in Line 6 of Algorithm 3.11 from RIVM into RIVMKF and providing the required function inputs R_t^1 and Q_t instead of λ.

Algorithm 3.19: Regularized IV M-Kalman filter (RIVMKF)

for $t \leftarrow 1$ to m do
  input: $\widehat{X}_{t-1}, P_{t-1}, A_t, B_t, \mathfrak{A}_t, W_{t,t}, R_t^1, Q_t, \kappa$
  $L_t = \left(W_{t,t} P_{t-1} \mathfrak{A}_t^\top\right)\left(R_t^1 + W_{t,t} A_t P_{t-1} \mathfrak{A}_t^\top\right)^{-1}$
  $P_t = (I - L_t A_t) P_{t-1} + Q_t$
  $P_t = P_t (I + \kappa P_t)^{-1}$
  switch *type* do
    case *Levenberg-Marquardt regularization* do
      $\widehat{X}_t = \widehat{X}_{t-1} + L_t\left(B_t - A_t \widehat{X}_{t-1}\right)$
    case *Tikhonov regularization* do
      $\widehat{X}_t = \widehat{X}_{t-1} + L_t\left(B_t - A_t \widehat{X}_{t-1}\right) - P_t \kappa \widehat{X}_{t-1}$
  output: $\widehat{X}_t, P_t$

Algorithm 3.20: Stenlund-Gustafsson IV M-Kalman filter (SGIVMKF)

for $t \leftarrow 1$ to m do
  input: $\widehat{X}_{t-1}, P_{t-1}, A_t, B_t, \mathfrak{A}_t, W_{t,t}, R_t^1, P_\mathrm{d}$
  $L_t = \left(W_{t,t} P_{t-1} \mathfrak{A}_t^\top\right)\left(R_t^1 + W_{t,t} A_t P_{t-1} \mathfrak{A}_t^\top\right)^{-1}$
  $\widehat{X}_t = \widehat{X}_{t-1} + L_t\left(B_t - A_t \widehat{X}_{t-1}\right)$
  $Q_t = \dfrac{W_{t,t} P_\mathrm{d} \mathfrak{A}_t^\top A_t P_\mathrm{d}}{R_t^1 + W_{t,t} A_t P_\mathrm{d} \mathfrak{A}_t^\top}$
  $P_t = (I - L_t A_t) P_{t-1} + Q_t$
  output: $\widehat{X}_t, P_t$

3.8.5 Stenlund-Gustafsson IV M-Kalman filter

Finally, the Stenlund-Gustafsson IV M-Kalman filter (SGIVMKF) is presented in Algorithm 3.20 and gives the last algorithm for the linear random-walk errors-in-variables problem. Please follow the steps which were described in Section 3.8.4 to build a SGIVMKF with robust estimated scale.

3.8.6 Experiments

Let us conduct the following experiment for the linear random-walk errors-in-variables model. This experiment grounds on Section 3.5.6 with the noise covariance matrix (3.46). However, in order to study different estimators for the random-walk errors-in-variables model (3.50), we will alter $\overline{X}_3$, which was adjusted to 3 in all previous experiments. Now, $\overline{X}_3$ depends on t and is realized as random-walk with

$$\overline{X}_{3,t} = \overline{X}_{3,t=1} + \sum_{i=1}^{t} \mathcal{N}(\mu, \sigma^2) \text{ with}$$

$$t = \begin{bmatrix} 1 & 2 & \dots & m \end{bmatrix}^\top, \; m = 10\,000\,\mathrm{s}$$

$$\mu = 0, \; \sigma = 1 \times 10^{-4}, \; \overline{X}_{3,t=1} = 3.$$

Experiment 3.6
Call the function linearRWeiv() two times with the strings 'Levenberg-Marquardt-RRIVM' and 'Levenberg-Marquardt-RIVMKF' to compute the two parameter estimates in Figure 3.15. Algorithm 3.11 with Algorithm 3.12 in Line 6 (RRIVM with robust estimated scale) will be executed with $\lambda = 0.995$ and $\kappa = 0.1$ for the first function call. The second function call runs Algorithm 3.19 inside Line 6 of Algorithm 3.11 (RIVMKF with robust estimated scale), where the covariance of parameter correction was set to $\mathrm{diag}(Q) = \begin{bmatrix} 1 \times 10^{-6} & 1 \times 10^{-8} & 1 \times 10^{-5} \end{bmatrix}$. The covariance of output noise was adjusted for both estimators to $R^1 = 0.02$, which meets the last entry of $\widetilde{P}$ in (3.46). The influence function was (3.15b) and the instruments were $\mathfrak{A}_t = A_{t-2}$ for both function calls.

Although RRIVM produces accurate results for $\widehat{X}_1$ and $\widehat{X}_2$ in Figure 3.15a, which was expected from the results of Figure 3.14b, this estimator tracks $\widehat{X}_3$ not precise. A delay between $\overline{X}_3$ and $\widehat{X}_3$ is clear to see in Figure 3.15a. A smaller forgetting factor would reduce this delay. However, the other estimates of $\widehat{X}_1$ and $\widehat{X}_2$ would show larger variation then.

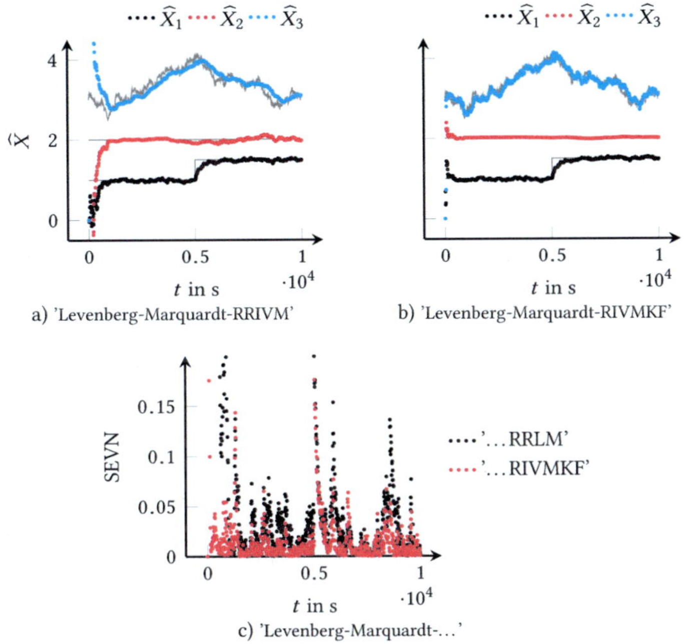

Figure 3.15: Estimated parameters for Levenberg-Marquardt-RRIVM and Levenberg-Marquardt-RIVMKF in Figure 3.15a and Figure 3.15b, respectively. Nominal values are shown with — lines. Figure 3.15c gives the Squared error vector norm.

There is a dilemma in recursive estimators with exponentially forgetting. A large forgetting factor reduces the parameter variance but then the algorithms loose tracking capability. One solution is to replace the time-invariant forgetting factor with a time-varying forgetting factor. This time-varying forgetting factor is then called variable forgetting and a vast of algorithms (sometimes ad hoc methods) have been presented [5, 56, 97, 123, 163, 170]. Another solution is to add perturbation to the P update [87]. Also methods which involve resetting of P are common.

However, all variable forgetting methods can be replaced by Kalman filter-based estimators if R^1 becomes time-varying. However, even if R^1 is time-invariant, as it is within this experiment, RIVMKF shows better performance in tracking of the random-walk like varying $\overline{X}_3$ in Figure 3.15b than RRIVM in Figure 3.15a. The superior accuracy of RIVMKF can also be seen in Figure 3.15c. This result was expected if we remember Section 3.7.1, where it was said that the Kalman filter is the optimal estimator to solve the random-walk output-error model, or the related random-walk errors-in-variables through the instrumental variables method. Moreover, remember that the Kalman filter covers RLS as special case. Hence, Kalman filter-based estimators offer the most flexibility and do not suffer under the λ-dilemma.

3.9 Linear polynomial-function output-error model

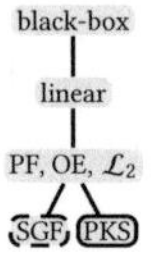

Contrary to Section 3.5–Section 3.8, this section covers a black-box model that we will use in Section 5.3 to smooth vehicle signals.

Reused material: Polynomial-functions have been shown suitable for extracting signals from noisy time series [148]. They are able to preserve the original signal level and have a well-defined and tunable delay.

The Savitzky Golay filter (SGF) uses convolution arrays to obtain the smoothed signal and the smoothed derivatives at one preselected point. Like proposed in [148], we use a smoothing window with equal left window (w_l) and right window (w_r) and model the signal with a time-varying polynomial-function of order $n-1$. SGF performs a weighted sum of the measured signal within the smoothing window. Hence, SGF requires one buffer of size $w_\mathrm{l}+1+w_\mathrm{r}$ for the measured data within the smoothing window and at least one buffer of same size for the weights which were derived in [104]. Additional buffers with specific weights are required as the number of desired derivatives increases.

3.9.1 Polynomial Kalman smoother

We use a specific form of the Kalman filter, which we call polynomial Kalman smoother (PKS), to extract noise from noisy measurements. PKS is based on the principles of SGF. The herein proposed PKS approach can be seen as the recursive version of SGF with exponentially weighted data. Conversely to SGF, PKS yields estimates of the polynomial-function parameters. Hence, we can evaluate the polynomial-function as well as its derivatives at multiple points. Due to the recursive approach, PKS outperforms SGF significantly in matters of memory.

The requirement for the polynomial-function approach is that the time series is built from auto-correlated signals. The following procedure would fail if the signal is a random process, because the time-varying polynomial-function would not properly model the underlying signal from the measured signal in this case.

[...]

We use the state-space representation in (3.52) with the state matrix ($\mathcal{A}$), the polynomial parameters X and the output matrix (C) as polynomial control input vector and perform a random-walk model [102] of the time-varying polynomial-function, whereas $\mathcal{B}$ and $\mathcal{D}$ are zero.

$$X_t = \mathcal{A} X_{t-1} + \mathcal{B} A_t \tag{3.52a}$$

$$B_t = C X_t + \mathcal{D} A_t \tag{3.52b}$$

For a fixed unit shift of the polynomial-function, the state matrix $\mathcal{A} \in \mathbb{Z}^{n \times n}$ becomes a time-invariant upper triangular square matrix and contains binomial coefficients $\binom{n}{k}$

$$\mathcal{A}_{i,j} = \begin{cases} \binom{j-1}{j-i} & \forall j \geq i \\ 0 & \forall j < i, \end{cases}$$

see [198] for general shifts. For instance, $\mathcal{A}$ yields for a polynomial-function with four parameters (third order) to

$$\mathcal{A} = \begin{bmatrix} 1 & 1 & 1 & 1 \\ 0 & 1 & 2 & 3 \\ 0 & 0 & 1 & 3 \\ 0 & 0 & 0 & 1 \end{bmatrix}.$$

If PKS is used as smoother with $w_\mathrm{l} = w_\mathrm{r}$, C is given with

$$C = \begin{bmatrix} (w_\mathrm{l} + 1 + w_\mathrm{r})^0 & (w_\mathrm{l} + 1 + w_\mathrm{r})^1 & \dots & (w_\mathrm{l} + 1 + w_\mathrm{r})^{n-1} \end{bmatrix}.$$

The PKS algorithm is shown in Algorithm 3.21, with the noisy measurement B_t, and the forgetting factor (λ). The smoothed signal at the center of the window ($w_\mathrm{l} + 1$) is gained by

$$\widehat{B}_{t-w_\mathrm{r}} = \begin{bmatrix} (w_\mathrm{l} + 1)^0 & (w_\mathrm{l} + 1)^1 & \dots & (w_\mathrm{l} + 1)^{n-1} \end{bmatrix} \widehat{X}_t$$

Algorithm 3.21: Polynomial Kalman smoother (PKS)

for $t \leftarrow 1$ to m do
- input: $\widehat{X}_{t-1}, P_{t-1}, B_t, \mathcal{A}, C, \lambda$
- $\widehat{X}'_{t-1} = \mathcal{A}\widehat{X}_{t-1}$
- $P'_{t-1} = \mathcal{A}P_{t-1}\mathcal{A}^\top$
- $L_t = \left(P'_{t-1}C^\top\right)\left(\lambda + CP'_{t-1}C^\top\right)^{-1}$
- $P_t = \frac{1}{\lambda}\left(P'_{t-1} - L_tCP'_{t-1}\right)$
- $\widehat{X}_t = \widehat{X}'_{t-1} + L_t\left(B_t - C\widehat{X}'_{t-1}\right)$
- output: $\widehat{X}_t, P_t$

3.9.2 Experiments

Let us compare PKS with an moving average finite impulse response (FIR) filter in the following signal smoothing experiment. We reuse B from Section 3.5.6 but omit the outliers in B, because PKS is a non-robust estimator. However, ✎ the derivation of robust PKS is similar to MKF in Section 3.7.2 and will be studied in future work. Another interesting topic would be a study about an errors-in-variables PKS (and it's robust sibling) which is similar to IVKF in Section 3.8.2.

> Experiment 3.7
> Call the function linearPFoe() to compute the estimated output and MSE in Figure 3.16. Algorithm 3.21 will be executed with $\lambda = 0.9$, $w_l = w_r = 20$ and a polynomial-function of fourth order (five parameters). FIR is realized with equally weighted 41 taps ($w_l + w_r + 1$) and computes the moving average of the measured output. The estimated output of PKS and FIR is delayed with w_r to bring the measured output and estimated output in phase.

Figure 3.16a presents a widely known result for moving average smoothed signals (herein we used a FIR filter with equal tap weights). FIR lacks under frequency dependent amplitude and phase response. The FIR smoothed estimated output masks the high frequent information of the underlying true output and it is difficult to compensate the time-varying phase shift.

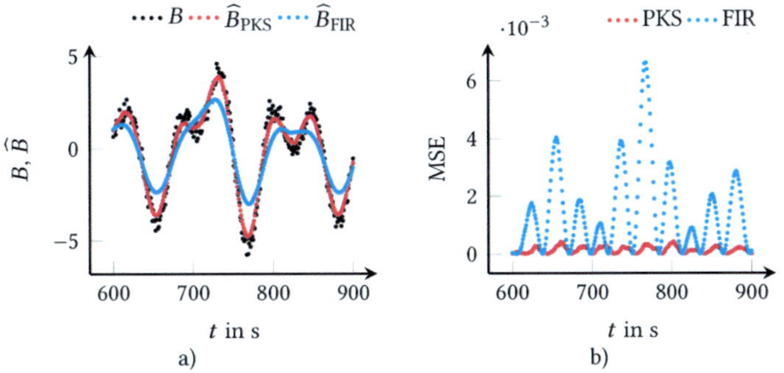

Figure 3.16: Figure 3.16a presents the estimated output of PKS and FIR (that was adjusted as moving average filter) with equal left window and right window. Note that the estimated output is delayed with w_r to bring B and $\widehat{B}$ in phase. Figure 3.16b gives the MSE of both estimators.

Although these drawbacks are widely reported in literature, moving average signal filtering is still popular.
PKS outperforms FIR largely. That is explainable due to the better flexibility of a polynomial-function with fourth order compared with moving average, which is indeed a polynomial-function of order zero. More precisely, PKS covers moving average filters by adjusting the order of the polynomial-function to zero.
Figure 3.16b shows the MSE for PKS and FIR for the same time-period. The MSE of PKS is generally lower and throughout more constant than the MSE of FIR. In conclusion, PKS outperforms FIR (in a moving average setting) by far in this signal smoothing experiment.
For ✎ the purpose of brevity, this section presented only an excerpt of possible PKS settings. Further research is required to compare PKS with other state of the art signal filters, because the presented moving average FIR is the simplest kind of signal filter and provides rather a weak benchmark.

3.10 Additional topics

This section covers the topics parameter constraints and model uncertainty. Both topics can be applied to all presented estimators (Section 3.5–Section 3.9). However, involving parameter constraints (Section 3.10.1) may be a challenge for black-box models, such as the presented polynomial-function output-error model in Section 3.9, because it is commonly hard to adjust parameter constraints when the parameters do not carry a physical meaning.

3.10.1 Constrained estimators

Remember that gray-box models differ from black-box models by the amount of prior knowledge, as explained in the introduction of this chapter (Page 33). As gray-box models ground mostly on physical laws, we will

have mostly a rough idea about the range of several parameters or can derive at least trivial inequality constraints, for instance $\widehat{X}_i \geq 0$.
Hence, we should consider constrained estimators, where the simplest one is parameter projection [65, pp. 91–94] and [126, pp. 52, 109, 112]. Parameter projection reduces into a simple parameter saturator if we impose individual lower bounds (X_{min}) and upper bounds (X_{max}) for each parameter [3, 174].
The general parameter saturator becomes

$$\widehat{X_{\text{c},i}} = \begin{cases} X_{\text{min},i} & \widehat{X}_i < X_{\text{min},i} \\ \widehat{X}_i & X_{\text{min},i} \leq \widehat{X}_i \leq X_{\text{max},i} \\ X_{\text{max},i} & \widehat{X}_i > X_{\text{max},i} \end{cases} \qquad \forall i$$

and may be used as subsequent step in conjunction with each estimator of Section 3.5–Section 3.9. Moreover [33, 174] and the references in the latter provide numerous examples in which imposed equality constraints improve the accuracy dramatically.
However, I like to invite the reader to study [162, pp. 212–223] and [161], who give an exceptional survey of constrained Kalman filters (inequality and equality constraints), and [7, 152] for constrained TLS estimators. A deeper discussion of parameter constraints, or constrained estimators in general, would exceed the scope of this dissertation.

3.10.2 Model uncertainty

Model uncertainty was briefly mentioned in Section 3.8.1 when robust Kalman filters were proposed as possible solution for the random-walk errors-in-variables model. Within this section, we will discuss another kind of model uncertainty. Specifically, we will discuss solutions for the case that we are uncertain how to adjust (or tune) λ or R^1 and Q.
Remember that λ was introduced for RLS in Section 3.5.3, whereas R^1 and Q were introduced for the Kalman filter in Section 3.7. Let us further focus

on the Kalman filter (and hence on R^1 and Q) because we know from (3.49) that the Kalman filter covers RLS as special case.
Upon here, it was implicitly assumed that prior knowledge about R^1 and Q is available. However, this prior knowledge is almost never available in practice. The common way is to interpret R^1 and Q as tunable values. Given enough experimental data, we adjust R^1 and Q empirically until some performance index of Section 3.2.1 is satisfactory. Further, we conduct model validation (preferably done by cross-validation, see Section 3.2.5) to become confident enough that the examined model will work for it's intended purpose. We will exactly follow this empirical approach in Chapter 5.
However, empirically tuned R^1 and Q have several drawbacks. First, this empirical approach requires expert knowledge. Second, this approach is time consuming, because we have to process large data, examine and interpret the results, and adjust R^1 and Q for the next loop. Third, the implicit assumption R^1 and Q being time-invariant may not hold.
Imagine that the accuracy of sensors may vary due to changing environmental conditions, such as vibration, temperature, luminance or simple the age of the sensor. Hence, R^1 and Q become time-varying, which was already indicated by the index t (R^1_t and Q_t) in Section 3.7.
In conclusion, when we tune R^1 and Q empirically over large experimental data, we will gain after several iterations reliable results for R^1 and Q, but these results will be an average for the given experimental data. Therefore, we will end up in a dilemma. As R^1 and Q depend on the experimental data, we seek for large data. However, gathering and processing of experimental data is time consuming and therefore costly.
Hence, ✎ let us briefly outline three alternative approaches to determine time-varying R^1_t and Q_t that appear attractive for further research.
First, Mehra [113] pioneered innovation-based adaptive estimation (IAE), given the fact that the Kalman filter innovations ($\widehat{B}$) become a sequel of white noise if R^1 and Q are set to their optimal values. Mehra [113] and most of the subsequent methods [4, 9, 16, 32, 115, 117, 120, 121, 132, 178, 194] require

large sliding windows for accurate covariance estimation. The number of references show that IAE is a broad field. Hence, let me recommend to consult the surveys in the introductions of [115, 121, 178] and the more educationally written sources [162, pp. 298–301] and [25] for an introduction in IAE.

Second, multiple model estimation (MME) comprises methods were individual tuned Kalman filters (each Kalman filter has a different setting for R^1 and Q) operate in parallel. Given the Bayesian probability for each Kalman filter (hence, for each model), MME delivers a probability weighted average over all models which is more accurate as each individual model. Static MME [162, pp. 301–305] and [15, pp. 441–443] converges to unity if the optimal model is in the set of candidate models. Otherwise, static MME converges to the model which is nearest to the optimal model. In case of time-varying systems, static MME would converge to a certain model and remain even if the system changes. Hence, dynamic MME methods and particularly interacting multiple model estimation (IMME) was invented [18] and [15, pp. 453–459]. IMME shows in many applications with model uncertainty convincing results [88, 169, 176, 177]. However, the bank of parallel Kalman filters causes more computational burden than IAE.

Third, Karasalo and Hu [89] propose an optimization approach for Q_t which does not require knowledge about the system dynamics. In other words, even $\mathcal{A}$ may be unknown. Moreover, [89] provide an excellent survey for IAE and MME methods in the introduction.

3.11 Which estimator should I use?

Throughout this chapter we studied numerous linear gray-box estimators (Section 3.5–Section 3.8) and one linear black-box estimator (Section 3.9). Let us define some general rules to answer the question of this section for linear gray-box estimators.

Choosing the right estimator depends upon our aim (what should the model do) and upon how certain we are in our assumptions about the system. In general, it is easier to define aims than assumptions and the following questions should help to find a proper estimator.

Question: Do I require highly accurate estimated parameters or do I need a good prediction quality from my model?

Errors-in-variables estimators are designed to reduce the parameter bias, whereas the purpose of output-error estimators is the minimization of the output-error. Further, random-walk errors-in-variables estimators reduce the parameter-bias further if the parameters vary on different rate, see Figure 3.15b. Indeed, random-walk errors-in-variables estimators are more general and include multi-input-single-output output-error and random-walk output-error estimators as special case. Output-error estimators suffer from biased parameters if certain assumptions (noise free measured inputs ...) do not hold, see Figure 3.14a. Hence, let us prefer random-walk errors-in-variables estimators for highly accurate estimated parameters.
For the purpose of a good prediction quality, multi-input-single-output output-error estimators work superior, see the discussion about the MSE at the end of Section 3.6.8.

Question: Show the measured inputs poor excitation at any time?

We know from Section 3.3 that poor excitation causes the wind-up problem, which results in unstable solutions for the estimated parameters, see Figure 3.8b. Further, as Hadamard's conditions cannot be ensured for recursive estimators, regularized estimators are worth to consider.

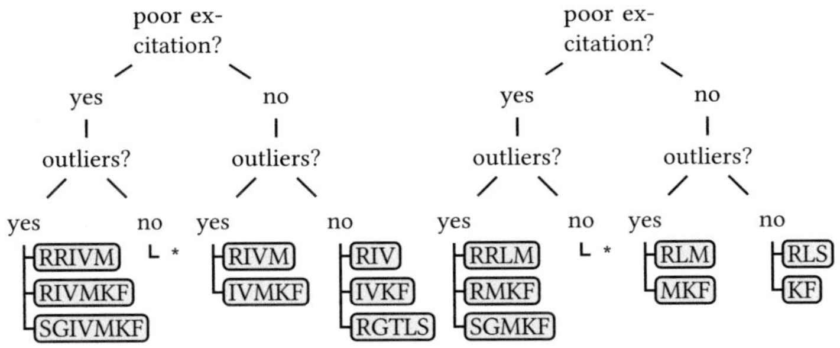

a) Highly accurate estimated parameters. b) Good prediction quality.

Figure 3.17: Decision tree for choosing the right estimator. Figure 3.17a lists estimators for highly accurate estimated parameters, whereas Figure 3.17b shows the respective estimators for good model prediction. *Regularized non-robust estimators are not explicitly presented herein. However, these methods can be derived from regularized and robust estimators.

Question: Do I expect outliers?

Remember that we defined in Section 3.4 outliers as observations which somehow deviate from our assumptions. If we are confident that the data is never corrupted by outliers, we may choose non-robust estimators. However, if we fail with this assumption, the consequences might be drastic, see Figure 3.4 and Figure 3.8a.

Hence, we should favor robust estimators although they cause larger computational burden and show slower convergence rate compared with non-robust estimators.

Figure 3.17 gives a decision tree to organize the presented recursive estimators of this chapter. First, decide between highly accurate estimated parameters (Figure 3.17a) or a good prediction quality (Figure 3.17b). Second, consider or neglect poor excitation. Third, think about outliers. The result is a group of recursive estimators which are appropriate for the intended application. Note that no-branches in Figure 3.17 give simple estimators, whereas yes-branches yield sophisticated methods.

Summary

This chapter provided a broad survey of linear gray-box models and their estimators. The presented connections and transitions between various models and estimators allowed us to derive more general estimators which include known basic estimators such as RLS as special case. Beside this, several margin notes marked open research topics for linear estimators.
Add to this, Section 3.9 introduced PKS as a novel estimator for polynomial-function black-box model. PKS outperformed FIR in a signal filtering experiment. Therefore, PKS will serve as signal filter for CAN signals in Section 5.3.
This survey prepared Chapter 4 which discusses related research and shows open research topics in vehicle parameter estimation. We emphasized recursive robust and regularized estimators which showed in several reproducible examples with increasing complexity superior performance compared with basic estimators.
The guidelines for estimator selection will support us in Chapter 5, where we solve real world problems in vehicle parameter estimation.

4 Survey of related research

Outline

The following state of research focuses on model-based estimators of vehicle parameters. We will discuss open problems and classify the state of research according to the model structure and estimator categories of Chapter 3.

4.1 Related research

In the last ten years a variety of methods to identify vehicle parameters have been introduced. These methods can be classified in recursive and batch. Thereby, recursive methods are characterized by a low calculation effort and are examined in the following. Former approaches mostly focus on few or even single parameter, whereby especially the vehicle mass is subject of extensive research work. Table 4.1 lists works, which deal with the identification of vehicle mass, coefficient of rolling resistance, longitudinal drag coefficient, and the road angle. The majority of the work uses the RLS procedure in combination with linear models of longitudinal vehicle dynamics. Two references use non-linear vehicle dynamics models, which require non-linear estimators such as the extended Kalman filter (EKF) and the unscented Kalman filter (UKF).
Vahidi, Stefanopoulou, and Peng [180] (and [181] for the journal version of [180]) used recursive least squares with multiple forgetting (RLSmf), which is a specific kind of RLS, to determine vehicle mass and road angle of a heavy duty vehicle. The rolling resistance and longitudinal aerodynamic resistance were additionally simulated and the longitudinal drag coeffi-

cient and coefficient of rolling resistance adjusted to achieve a satisfactory goodness of fit between measurement and model. Thereby, the number of unknown parameters was reduced, but this approach is arbitrarily, because the longitudinal drag coefficient and coefficient of rolling resistance are only roughly known in practice and subject to variation due to environmental changes. The longitudinal vehicle dynamics model proposed by [180, 181] was

$$\dot{v}_{\mathrm{V}} = \left(\frac{T_{\mathrm{E}} - I_{\mathrm{E}} \ddot{\theta}_{\mathrm{E}}}{r_{\mathrm{W}}/(i_{\mathrm{G}} i_{\mathrm{D}})} - \frac{T_{\mathrm{B}}}{r_{\mathrm{W}}} - \frac{\rho_{\mathrm{a}}}{2} A_{\mathrm{V}} c_x v_{\mathrm{V}}^2 \right) \frac{1}{m_{\mathrm{V}}} - \frac{\sin\left(\theta_{\mathrm{r}} + \arctan f_{\mathrm{r0}}\right)}{\cos\left(\arctan f_{\mathrm{r0}}\right)} g. \quad (4.1)$$

Table 4.1: Current state of research for model-based identification of external resistances and forces on vehicles. The specific resistances and forces are the climbing force (F_{cl}), acceleration force (F_{ac}), longitudinal aerodynamic resistance ($F_{x_{\mathrm{a}}}$), and the rolling resistance (F_{Wr}).

Reference	Model & estimator	Force				Parameter
		F_{cl}	F_{ac}	$F_{x_{\mathrm{a}}}$	F_{Wr}	
[180, 181]	(4.1) RLSmf	✓	✓	✗	✗	$m_{\mathrm{V}}, \theta_{\mathrm{r}}$
[112]	(4.1) RLS	✓	✓	✗	✗	$m_{\mathrm{V}}, \theta_{\mathrm{r}}$
[193]	(4.1) EKF	✓	✓	✗	✗	$m_{\mathrm{V}}, \theta_{\mathrm{r}}$
[52]	(4.2) RLS	✗	✓	✗	✗	m_{V}
[13]	(4.3) RLS	✓	✓	(✓)	✓	$m_{\mathrm{V}}, \theta_{\mathrm{r}}, (c_x)^1, F_{\mathrm{Wr}}$
[196]	(4.5) RLSmf	(✓)[2]	✓	✓	✓	$m_{\mathrm{V}}, (F_{x_{\mathrm{a}}} + F_{\mathrm{Wr}})$
[71, 83]	(4.6) RLS, KF, ad-hoc MME[3]	✓	✓	✓	✓	$m_{\mathrm{V}}, (c_x)^{1,4}, (F_{\mathrm{cl}} + F_{\mathrm{Wr}})^4$
[70]	(4.7) RLS	✗	✓	✓	✓	m_{V}
[78, 79]	(4.8) EKF,UKF	✗	✓	✓	✓	$m_{\mathrm{V}}^{*\,5}, I_{z\mathrm{V}}$

[1] c_x is hidden in a lumped Parameter $\frac{\rho_{\mathrm{a}}}{2} A_{\mathrm{V}} c_x$.
[2] However, the test were performed on a flat road.
[3] Not comparable with static MME or dynamic MME (IMME) of see Section 3.10.2.
[4] No results presented.
[5] m_{V}^* denotes the vehicle sprung mass.

Note that (4.1) does neither consider reduced moment of inertia nor friction losses inside the drive-train. However, Vahidi, Stefanopoulou, and Peng [181] stated that T_{E} can be multiplied with an appropriate coefficient of efficiency to account for drive-train losses. According to [193], the derivation of (4.1) is based on the addition theorem

$$\sin(x \pm y) = \sin x \cos y \pm \cos x \sin y,$$
$$\frac{\sin(x + y)}{\cos y} = \sin x + \cos x \tan y.$$

The last term in (4.1) is equivalent to $g(f_{\mathrm{r0}} \cos \theta_{\mathrm{r}} + \sin \theta_{\mathrm{r}})$ and allows to separate the road angle. The two unknown lumped parameters are $\sin(\theta_{\mathrm{r}} + \arctan f_{\mathrm{r0}})$ and $1/m_{\mathrm{V}}$. The derived RLSmf in [180, 181] is an attempt to solve the linear random-walk output-error model (3.48). However, the relation between RLS and the Kalman filter (which was derived for state estimation but can also be used for parameter estimation) was not used in the development of RLSmf, and RLSmf relies on the restrictive assumption that there is no correlation between the individual parameters. The derivation of RLSmf in [180, 181] shows that this estimator solves indeed two separate and independent single-input-single-output (SISO) models with parallel and independent RLS estimators. Basically, the result of the first estimator from the previous time step is substituted into the cost function of the second estimator. Hence, RLSmf is a kind of decoupled estimator. However, the experiments in [135] have shown that RLSmf diverges if the individual parameters show correlation. Vahidi, Stefanopoulou, and Peng [180] emphasize several times that rich excitation is important to obtain precise estimates. Specifically, Vahidi, Stefanopoulou, and Peng [180] state that:

> For successful identification we made sure that the dynamics is sufficiently rich, many times by asking the driver to pulse the commands like throttle and braking.

The same comment gives [181, p. 35].

McIntyre et al. [112] present a two step approach to identify the vehicle mass and road angle of heavy duty vehicle with a vehicle mass of about 20 t with model (4.1). First, the vehicle mass and the road angle were estimated with a RLS procedure. Thereby, the road angle was assumed to be constant. Second, the road angle was estimated more accurately with a non-linear estimator. Depending on the setup, the error of the vehicle mass is smaller than ±5 %. Similar to [180, 181], McIntyre et al. [112] mention that rich excitation is required for accurate estimates:

> Persistent excitation is required for the accurate estimation of parameters, which was enforced by choosing a sufficiently varying fueling profile.

Winstead and Kolmanovsky [193] identify the same parameters simultaneously with an EKF for a passenger car. Here a speed cruise controller (model predictive control), which optimizes the velocity trajectory of the vehicle in order to ensure sufficient rich excitation, reduces the variance of the parameter estimations. Additionally, a low divergence between the actual and the nominal velocity was assured. Here (4.1) was modified by a power-train efficiency factor and a friction term, which was not specified any further.
Fathy, Kang, and Stein [52] focused on the vehicle mass. High-frequency parts of longitudinal vehicle dynamics were allocated to the acceleration force. This allocation allows an estimation of the vehicle mass without the knowledge of the remaining low-frequency parts of the external forces by

$$m_{\mathrm{V}}\dot{v}_{\mathrm{V}} = \frac{T_{\mathrm{R}} - T_{\mathrm{B}}}{r_{\mathrm{W}}}. \tag{4.2}$$

RLS was the applied vehicle mass estimator. A band-pass filter separated the high- and the low-frequency parts and eliminated high-frequent noise. A fuzzy logic determined driving conditions which were primarily longitudinal dynamic. The convergence of the proposed method requires rich excitation [52, p. 1847].

Bae, Ryu, and Gerdes [13] measured the road angle with a global positioning system. Estimates for the vehicle mass, the rolling resistance, and the longitudinal drag coefficient of a passenger car were determined by

$$\frac{T_\mathrm{R}}{r_\mathrm{W}} = m_\mathrm{V}(\dot{v}_\mathrm{V} + g \sin\theta_\mathrm{r}) + \frac{\rho_\mathrm{a}}{2} A_\mathrm{V} v_\mathrm{V}^2 c_x + F_\mathrm{Wr} \tag{4.3}$$

with RLS. Note that the longitudinal drag coefficient was lumped with the air density and the vehicle cross-sectional area and is therefore hidden. The vehicle mass converges with an accuracy of ±2 %. However, the lumped parameter, which covers the longitudinal drag coefficient, shows a large divergence of the expected value. The driven velocity profile [13, Fig.7] shows a serrated course which assured rich longitudinal dynamics (rich excitation) within the measurement. Bae, Ryu, and Gerdes [13] justified the velocity profile with:

> A mix of acceleration of the vehicle followed by deceleration (letting the accelerator pedal up without engaging foot brake pedal) was repeated to simulate real world situations and generate excitation for judging the stability of the estimate.

However, I am afraid I have to disagree that the presented velocity profile represents many real world situations. Although the discussion of various *driving styles*, their classification, and driver comfort is beyond this dissertation, continuous acceleration and deceleration is a sportive driving style which may not be ensured at all times.

Yu et al. [196] presented a vehicle mass estimator with an accuracy of ±2.8 % for an experimental electric vehicle. The used vehicle longitudinal dynamics model was

$$m_\mathrm{V}\dot{v}_\mathrm{V} = \sum_{j_\mathrm{W}=1}^{4}\left(\frac{T_{\mathrm{R},j_\mathrm{W}} - I_{\mathrm{W},j_\mathrm{W}}\ddot{\theta}_{\mathrm{W},j_\mathrm{W}}}{r_\mathrm{W}}\right) - \dots$$

$$\dots - f_{\mathrm{r}0} m_\mathrm{V} g \cos\theta_\mathrm{r} - m_\mathrm{V} g \sin\theta_\mathrm{r} - \frac{1}{1.63} c_x A_\mathrm{V} v_\mathrm{V}^2,$$

which was further simplified with the relation

$$\ddot{x}_{\text{VCAN}} = \dot{v}_{\text{V}} + g \sin\theta_{\text{r}}, \tag{4.4}$$

into

$$\sum_{j_{\text{W}}=1}^{4} \left(\frac{T_{\text{R},j_{\text{W}}} - I_{\text{W},j_{\text{W}}} \ddot{\theta}_{\text{W},j_{\text{W}}}}{r_{\text{W}}} \right) = m_{\text{V}} \ddot{x}_{\text{VCAN}} + \dots$$

$$\dots + f_{\text{r0}} m_{\text{V}} g \cos\theta_{\text{r}} + \frac{1}{1.63} c_x A_{\text{V}} v_{\text{V}}^2. \tag{4.5}$$

Here, $\ddot{x}_{\text{VCAN}}$ is the vehicle acceleration which is measured by a sensor. This signal is corrupted by the road angle and gravitational constant. The last two terms in (4.5) (in the second line) were treated as lumped parameter and RLSmf was used as estimator.

Huh et al. [83] presented a vehicle mass estimator which consists of three independent models for vehicle longitudinal, lateral, and vertical dynamics. RLS was used for the vehicle longitudinal and vertical model, whereas the Kalman filter gave estimates for the vehicle lateral model. The vehicle longitudinal dynamics model was

$$\frac{T_{\text{R}}}{r_{\text{W}}} = \dot{v}_{\text{V}} m_{\text{V}} + \frac{\rho_{\text{a}}}{2} A_{\text{V}} v_{\text{V}}^2 c_x + g\big(f_{\text{r0}} m_{\text{V}} \cos\theta_{\text{r}} + m_{\text{V}} \sin\theta_{\text{r}}\big), \tag{4.6}$$

in which the last two terms were treated as lumped parameter, as well as $\frac{\rho_{\text{a}}}{2} A_{\text{V}} c_x$. Note that model (4.6) is similar to model (4.5). Given thresholds for v_{V}, $\dot{v}_{\text{V}}$, $\dot{\psi}_{\text{V}}$, and $\ddot{z}_{\text{V}}$ a Boolean logic classifies the current driving style into longitudinal, lateral, or vertical. Given the classified driving style, an additional RLS estimator updated the unified estimate for the vehicle mass with the current result from the respective model. Hence, the method in [83] is some kind of multiple model estimation (MME), that grounds on a simple classifier and unifies the result of each model with RLS. However, [83] is an ad hoc method and the result for the unified vehicle mass estimate

should strongly depend on the classifier thresholds. Hence, [83] is not comparable with more sophisticated static MME or dynamic MME (IMME), see Section 3.10.2. Moreover, it is likely that the used ad hoc classifier is obsolete by using a MME method. All results were gained from simulations. The methods and models in [71] are similar to [83]. However Han et al. [71] omited the vehicle vertical dynamics model of [83], used (4.4) instead of $\dot{v}_V$ as measured input, and provide results from real world data. Due to the high similarity, [83] and [71] are merged in Table 4.1.
Halfmann and Holzmann [70, pp. 63–75] presented a RLS vehicle mass estimator which is also based on a longitudinal dynamics model

$$\dot{v}_V = \frac{1}{m_V}\left(F - \frac{\rho_a}{2} A_V v_V^2 c_x\right), \tag{4.7a}$$

$$F = F_{Wr} + \frac{T_R - I_W \ddot{\theta}_W}{r_W}. \tag{4.7b}$$

Note that (4.7) requires knowledge or measurements of all quantities apart from the vehicle mass. Highway trips between $70\,\mathrm{km\,h^{-1}}$ to $120\,\mathrm{km\,h^{-1}}$ were seen as favorable experimental condition and selected by a logic. The road angle was not considered. However, a road angle observer is presented later on in [70, pp. 182–189]. Similar to [13], the driven velocity profile [70, Figure 4.6] shows rich excitation.
Upon here, the overview of related research focused on longitudinal vehicle dynamics based parameter estimation. The introduction in [42] provides an excellent survey of vehicle mass estimation references which are based on lateral dynamics, drive-train dynamics, and suspension dynamics. Moreover, Mayer [109] presented sophisticated methods to measure certain vehicle parameters with additionally sensors.
Now we will discuss two references which are based on lateral dynamics. A non-linear double lane vehicle dynamics model with roll dynamics was used in [78, 79] to estimate the vehicle sprung mass and vehicle yaw moment of inertia (I_{xV}). The climbing force was not considered and assumed to

be negligible. The non-linear state-space model comprises the five states $\begin{bmatrix} \dot{x}_\mathrm{V} & \dot{y}_\mathrm{V} & \dot{\psi}_\mathrm{V} & \phi_\mathrm{V} & \dot{\phi}_\mathrm{V} \end{bmatrix}^\mathsf{T}$, which gives a four degree of freedom vehicle model. The parameter vector is

$$X = \begin{bmatrix} m_\mathrm{V}^* & I_{z\mathrm{V}} \end{bmatrix}^\mathsf{T} \tag{4.8a}$$

$$X = \begin{bmatrix} m_\mathrm{V}^* & l_{12} \end{bmatrix}^\mathsf{T} \tag{4.8b}$$

in [78] and [79], respectively, where m_V^* denotes the vehicle sprung mass. Note that the vehicle yaw moment of inertia in [79] is computed with the estimated sprung mass and the distance between front axle and vehicle center of gravity position l_{12}. Hong et al. [78] compared the performance of EKF and UKF through simulations and assumed that the vehicle states are given. The performance of UKF in terms of vehicle parameter estimation is superior compared with EKF. Hong et al. [79] considered the dual estimation problem through dual unscented Kalman filtering (dual UKF). The dual estimation problem occurs when states and parameters are coupled, which means that some states are required for parameter estimation which affects however the state estimation. The vehicle sprung mass estimation results from test drives on a track with two chattered ramps, which was explained by noise and unmodeled dynamics [79]. Therefore, Hong et al. [79] applied posteriori signal processing for the parameter estimates. This signal processing stabilized and smoothed the vehicle sprung mass and vehicle yaw moment of inertia estimates, which were highly precise.

Finally, let me briefly mention two own references which fit in this survey of related research. Rhode and Gauterin [134] proposed a RTLS estimator which is based on a recursive singular value decomposition update scheme and presented vehicle mass estimates. Rhode and Gauterin [135] generalized this RTLS estimator into RGTLS through data scaling and showed estimates for $f_{\mathrm{r}0}$, $f_{\mathrm{r}1}$, and $A_\mathrm{V} c_x$ in addition to the vehicle mass.

4.2 Conclusions

Most ✎ discussed references focus on vehicle mass estimation and many references rely on specific assumptions of the driving style. The vehicle tractive force is not accurate enough modeled in some references (some important force components of the vehicle tractive force were neglected) and up to now, an integral approach for the prediction of the vehicle tractive force is missing. Further, the current research lacks missing model validation and model selection methods or any measure for the model's goodness of fit. Furthermore, most methods require a high rate of longitudinal dynamics (rich excitation) to gain exact estimation results. Regularization or other methods to avoid wind-up during poor excitation were not considered.

The majority of references use the linear multi-input-single-output output-error model of Section 3.5 together with the RLS estimator. RLSmf on the other hand, is an attempt to solve the linear random-walk output-error model of Section 3.7 with the aim to account for time-varying parameters with different variation rate. However, RLSmf suffers under restrictive assumptions and the convergence of RLSmf is neither studied nor ensured. Hence, I propose to treat RLSmf as ad hoc estimator which works well in specific applications and instead let us prefer the better studied Kalman filter as estimator for the random-walk output-error model.

Generally, the concept of robustness and methods to avoid biased parameters were not discussed although all references seek for highly precise parameter estimates. To the best of my knowledge, [134, 135] are the only references in vehicle parameter estimation which consider the multi-input-single-output errors-in-variables model. There is no reference for the the random-walk errors-in-variables model.

To sum up, the current state of research offers many open research topics and this dissertation attempts to address some of these open topics for vehicle parameter estimation and state prediction in Chapter 5.

Summary

The survey of related research in vehicle parameter estimation within this chapter highlighted open research topics and missing estimators which will be addressed in Chapter 5.

5 Vehicle tractive force prediction and mass estimation

> "Essentially, all models are wrong, but some are useful."
>
> *Box and Draper [20, p. 424].*

Outline

This chapter applies estimators from Chapter 3 to estimate parameters of gray-box models which ground on vehicle dynamics of Chapter 2. The primary goal is a precise prediction of the vehicle tractive force for look-ahead controllers, which are required in energy efficient driving or autonomous driving. The vehicle tractive force prediction quality of two gray-box models, each in combination with several estimators, is compared to the benchmark with a given white-box model on a large set of real world vehicle data. Additionally, vehicle mass estimation results will link this dissertation more closely to related research of Chapter 4.

5.1 Experimental conditions

The experiments were conducted with two grand touring sport cars from Porsche company. First, a Porsche Panamera two wheel rear drive vehicle with a 228 kW (310 horse power) 3.6 L V6 engine, 7-speed dual-clutch gearbox, and carbon dioxide emission of 196 g km^{-1} [129]. Second, a Porsche Panamera Turbo S four wheel drive vehicle with a 419 kW (560 horse power)

4.8 L V8 engine, 7-speed dual-clutch gearbox, and carbon dioxide emission of 239 g km^{-1} [128].

Each vehicle was equipped with the rapid prototyping real-time hardware dSpace MicroAutoBox II, which offers a 900 MHz IBM PowerPC, 16 MB storage, and Matlab Simulink integration [46]. All required CAN signals were recorded at 100 Hz with dSpace control desk software on a laptop.

The MicroAutoBox II was the only modification from series Panameras. Hence, only the original equipped sensors were available throughout the test runs. The most important sensors are the four wheel speed $\left(\dot{\theta}_{\mathrm{WCAN}}\right)$ sensors from the anti-lock braking system, the vehicle longitudinal ($\ddot{x}_{\mathrm{VCAN}}$), lateral acceleration ($\ddot{y}_{\mathrm{VCAN}}$) and yaw rate ($\dot{\psi}_{\mathrm{VCAN}}$) sensors from the electronic stability program, the steering wheel angle (δ_{SCAN}) sensor and the four braking pressure (p_{BCAN}) sensors from the anti-lock braking system. The precise sensor type cannot be given herein. However, [19] contains information about similar vehicle specific sensors.

The data set contains records of 85 test runs from the period November 2012 to April 2014. All test runs were conducted in Southern Germany on public roads, which were mostly hilly countryside roads with a few motorway sections. The overall distance is 1403 km. The shortest test run is fewer than 1 km and the longest 110 km on a motorway. The overall test time is 78 516 s (≈22 h). The shortest four test runs contain not enough data to split the data into training data and validation data for model validation. Hence, the active number of test runs reduces to 81.

During the cold weather period, 22 test runs were conducted with winter tires. We used the Panamera two wheel drive V6 vehicle for the majority of test runs (54 out of 81). The distribution of test runs per month is: 47 in June, 13 in March, 12 in November, 7 in April, and 2 in May. The weather conditions vary accordingly from cold, windy and wet to dry, warm weather. However, the environmental conditions were not recorded. Also the driving style varies broadly due to different drivers on different road types: city, countryside, and motorway. Also, the number of passengers varies (vehicle mass) as well as the use of air conditioning, window lifters, and sunroof (air drag).

The ✎ classification of all environmental conditions and driving styles is far beyond the scope of this dissertation. Hence, although the data contains a broad dispersion of experimental conditions, it is not proven that the data covers all use cases that may arise in practice. Herein, the principle was to collect as much data as was available for estimation and model validation. If we want to alter the experimental condition in a more systematic way, hardware in the loop could be a method of choice [61].

One big drawback of hardware in the loop compared with real world data is that hardware in the loop gives model-based validation data, because hardware in the loop is a model itself, but not the real world system. Even if we take an entire vehicle on a test bench, the environmental conditions needs to be modeled. Hence, outliers may not arise because they are not modeled. Anyhow, also hardware in the loop does not answer the question: *how much testing is enough testing?*, which is an own research topic and also not within the scope of this dissertation.

Figure 5.1 gives four characteristic vehicle states of one test run. Three challenges arise when working with real world data. First, apart from the gear in Figure 5.1, all presented vehicle states are not directly measurable through vehicle sensors. We will address this problem in Section 5.2. Second, real world data may be corrupted with missing data and probably outliers. The vehicle tractive force in Figure 5.1a shows missing data in several operational conditions. Third, some vehicle states show poor excitation which might end up in estimator wind-up, which then requires regularized or wind-up stable estimators (see Section 3.3). The vehicle tractive force in Figure 5.1a as well as the vehicle longitudinal acceleration in Figure 5.1b shows poor excitation for $t = 600\,\mathrm{s}$ to $640\,\mathrm{s}$. Hence, Figure 5.1 supports the conclusions in Section 4.2 which were made for Chapter 4 (Survey of related research). Precisely missing robust and regularized or wind-up stable estimators.

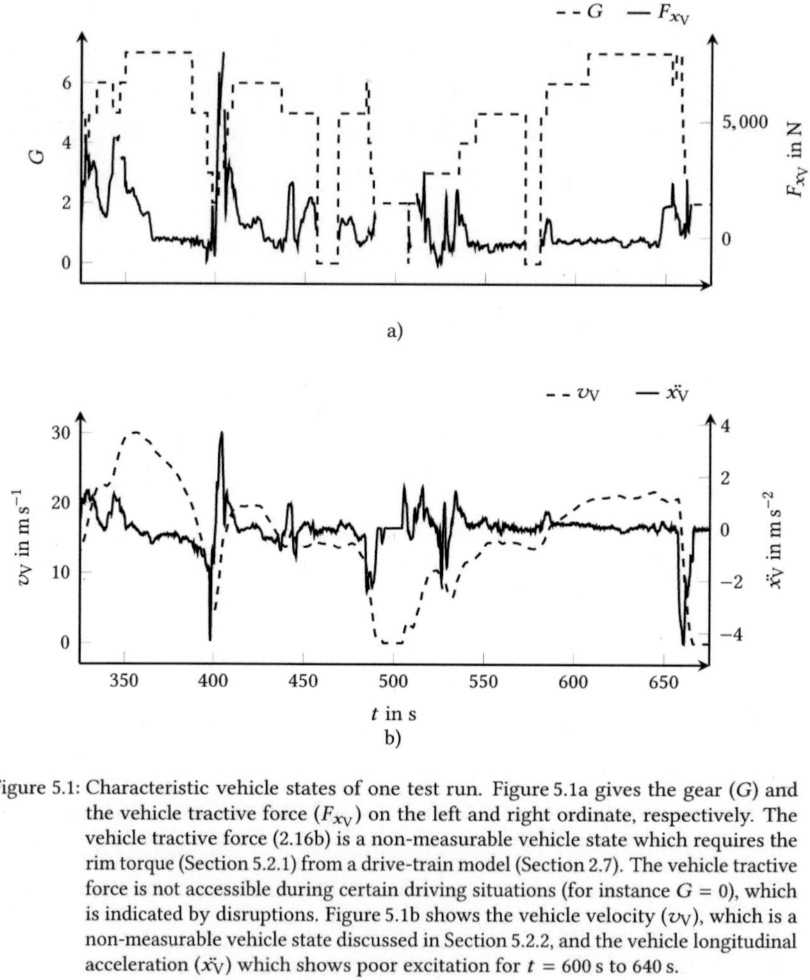

Figure 5.1: Characteristic vehicle states of one test run. Figure 5.1a gives the gear (G) and the vehicle tractive force (F_{xV}) on the left and right ordinate, respectively. The vehicle tractive force (2.16b) is a non-measurable vehicle state which requires the rim torque (Section 5.2.1) from a drive-train model (Section 2.7). The vehicle tractive force is not accessible during certain driving situations (for instance $G = 0$), which is indicated by disruptions. Figure 5.1b shows the vehicle velocity (v_V), which is a non-measurable vehicle state discussed in Section 5.2.2, and the vehicle longitudinal acceleration ($\ddot{x}_V$) which shows poor excitation for $t = 600$ s to 640 s.

5.2 Estimation of non-measured vehicle states

We discussed in Section 5.1 that the vehicles were purely equipped with original sensors which did not cover the full range of required vehicle states. Hence, we will introduce observers and estimators which yield non-measurable vehicle states from available sensor measurements.

5.2.1 Rim torque

The drive-train model of Section 2.7 is a mixed white-box and black-box model and gives the rim torque as function of the engine torque, gear, gear shaft speed, and reduced moment of inertia. All required inputs were supplied by control area network (CAN) data. However, only some of these inputs were measured through sensors. The engine torque comes from an engine model which uses lookup tables and sensor information (throttle position) [compare 155, pp. 217–219], whereas the reduced moment of inertia comes as function of the gear from a vehicle specific lookup table. Note that drive-train model has a rather simple model structure and is designed for steady state to low dynamics. High frequent drive-train dynamics would require to consider stiffness and damping for all shafts, mountings, and couplings [155, pp. 205–224]. Furthermore, transients such as time-varying oil temperature were neglected. Hence, it is likely that the drive-train model gives a biased rim torque until all parts of the drive-train operate at steady state temperature.

5.2.2 Vehicle velocity

The vehicle velocity is a major vehicle state which determines several subsequent vehicle states such as the wheel slip (and thereby the tire-road contact Section 2.2), and the longitudinal aerodynamic resistance of Section 2.4.3. An example for a direct vehicle velocity sensor is the Correvit sensor from Kistler [92]. However, such additional sensors are expensive and only used for model validation [85, p. 144].

The simplest estimator for the vehicle velocity becomes

$$v_V = \frac{r_W}{2}\left(\dot{\theta_{W3}} + \dot{\theta_{W4}}\right), \quad (5.1)$$

which is basically an average of the non-driven front wheel speeds. However, wheel slip and cornering (slip angle) is not considered in (5.1), which leads to biased estimates for the vehicle velocity.
Accordingly, more sophisticated vehicle velocity estimators were invented. Kiencke and Nielsen [91, pp. 351–363] give equations to correct the wheel speed data from cornering bias ([91, Figure 9.2]) and present two vehicle velocity estimators. First, a Kalman filter uses all four wheel speeds of the vehicle $\left(\dot{\theta_{W1,\ldots,4}}\right)$ and the longitudinal acceleration sensor ($\ddot{x_{VCAN}}$). The covariance of output noise is treated as tunable time-varying parameter which depends on the current driving style. Actually, this method is a simple and promising sensor fusion approach. Second, a Fuzzy Logic is presented, which might require more tuning but reduces the computational burden. Another rich source of Fuzzy Logic vehicle velocity estimation is [157, pp. 73–86] ([as book version 85, pp. 155–159]).
We used the existing CAN signal from the electronic stability program for the vehicle velocity. However, the precise definition of this CAN signal cannot be discussed herein.

5.2.3 Path angle

The gradient angle determines the climbing force (2.5). It is common to assume that gradient angle and road angle are equal, $\theta = \theta_r$, which is true for straight driving. However, we can easily design a counterexample. Imagine a zigzag driving style which results in a longer path on lower gradient angle, $\theta < \theta_r$. On the other hand, zigzag driving is quite limited due to the lane width. Hence, $\theta = \theta_r$ is a feasible assumption in practice.

Accordingly, road angle estimators were subject to extensive research [147], [91, pp. 402–408], [70, pp. 182–189], [175, pp. 130–132] and [157, p. 56]. Additionally, mixed road angle and vehicle mass estimators are given by [112, 180, 181, 193].

Let us look on a simple model for the gradient angle. The relation (4.4) leads to

$$\theta = \arcsin \frac{\ddot{x}_{\mathrm{VCAN}} - \dot{v}_{\mathrm{V}}}{g}, \tag{5.2}$$

which gives a gray-box model for the gradient angle. The term $\ddot{x}_{\mathrm{VCAN}}$ depicts the longitudinal acceleration sensor measurement from the body fixed acceleration sensor (compare Figure 2.1 for the definition of vehicle longitudinal axis), whereas $\dot{v}_{\mathrm{V}}$ is the derivative of the vehicle velocity from Section 5.2.2.

Trabelsi [175, p. 131] and Kiencke and Nielsen [91, p. 403] used (5.2) and filtered data to determine the road angle over time. Kiencke and Nielsen [91, p. 404] report an error of <5 % and Trabelsi [175, p. 132] an error <2 %. Semmler [157, p. 56] used (5.2) with a not further explained filter and highlighted in [157, Figure 5.6] delay between estimated road angle and validation data.

As the reported results in Trabelsi [175, p. 132], Kiencke and Nielsen [91, p. 404] and [157, Figure 5.6] of (5.2) are promising, let us apply the PKS from Section 3.9.1 with

$$\begin{aligned} B_t &= \arcsin \frac{\ddot{x}_{\mathrm{VCAN},t} - \dot{v}_{\mathrm{V},t}}{g} \\ \widehat{B}_t &\approx \theta_{,t}, \end{aligned}$$

to estimate the gradient angle recursively. Accordingly, the estimated output of PKS gives the gradient angle estimate of the noisy time series from model (5.2). Figure 5.2 compares the gradient angle estimate (given as path grade in %) of PKS with road angle validation data from predictive route data (PSD) [11, 119].

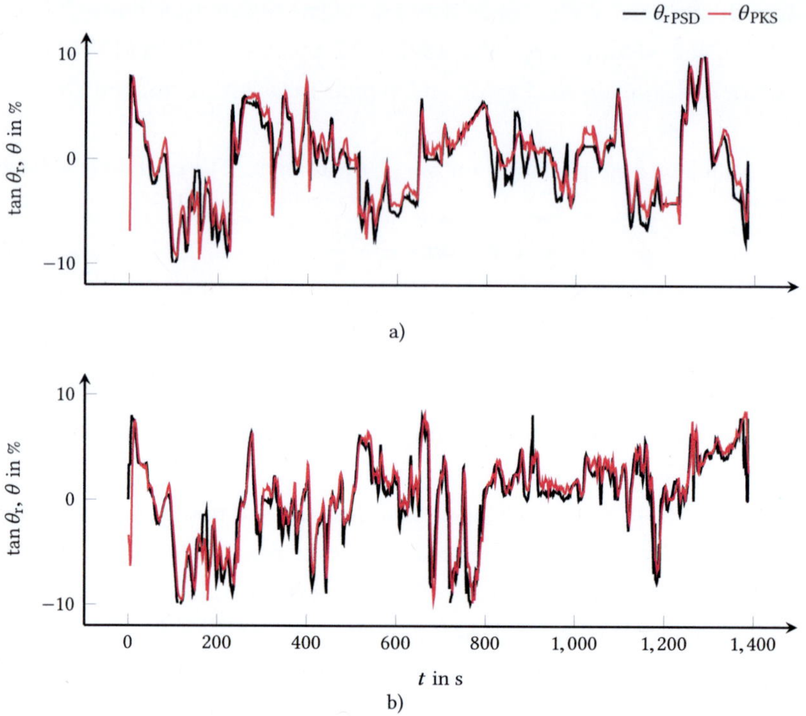

Figure 5.2: Path grade estimation with PKS for two independent test runs. The road grade validation data comes from predictive route data (PSD) [11, 119].

Predictive route data is part of advanced vehicle navigation systems and gives predictive road information, about: road grade, road curvature, road class, lane-width, and road surface. This information depends on the current vehicle position, planned route, and the most probable path at intersections. PKS and PSD show good correlation. However, PKS does not precisely follow the validation data in Figure 5.2a between t = 800 s to 1000 s. On the other hand, PSD contains outliers in Figure 5.2b for t = 920 s, 1060 s, and 1250 s. Hence, the slight imprecision of PKS in Figure 5.2a is rather maintainable compared with the outliers which PSD gives.

Note ✎ that (5.2) does not cover bias in the acceleration sensor measurement due to cornering (vehicle side slip angle), vehicle pitch angle, vehicle roll angle and road bank angle. Hence, the deviation between PKS and PSD in Figure 5.2a between $t = 800\,\mathrm{s}$ to $1000\,\mathrm{s}$ might be caused from specific driving situations which are not modeled by (5.2).
Isermann [85, p. 156] provides the required transformations to correct cornering, pitch, and roll bias from the longitudinal and lateral acceleration sensors.

$$
\begin{aligned}
\ddot{x_{\mathrm{V}}}' &= \ddot{x}_{\mathrm{VCAN}} + g \sin \theta_{\mathrm{V}} \cos \phi_{\mathrm{V}} \\
\ddot{y_{\mathrm{V}}}' &= \ddot{y}_{\mathrm{VCAN}} - g \sin \phi_{\mathrm{V}} \cos \theta_{\mathrm{V}} \\
\dot{v_{\mathrm{V}}} &= \ddot{x_{\mathrm{V}}}' \cos \beta_{\mathrm{V}} + \ddot{y_{\mathrm{V}}}' \sin \beta_{\mathrm{V}} - g \sin \theta
\end{aligned}
$$

However, the road bank angle is not considered and this topic might offer enough research questions for future work.
Add to this, inaccuracies in the vehicle velocity CAN signal could be the reason for the deviation between PKS and PSD.

5.2.4 Path radius

The path radius is required to describe lateral vehicle dynamics. Specifically, the computation of the cornering resistance in Section 2.5.1 needs the path radius. Moreover, lane keeping and illumination driver assistance systems need road geometry information. Accordingly, vision-based road curvature estimators were proposed in [39, 72, 190].
We had no access to vision signals. Hence, let us use the single track model of Figure 2.7 and write the Ackermann steer angle [155, p. 245] and [1]

$$
\tan \delta_{\mathrm{W}12} = \frac{l_{12} + l_{34}}{\sqrt{r^2 - l_{34}^2}},
$$

which can be rewritten to determine the path radius

$$r = \sqrt{\left(\frac{l_{12} + l_{34}}{\tan \delta_{W12}}\right)^2 + l_{34}^2}. \tag{5.3}$$

The wheel steer angle δ_{W12} of the front wheels was not directly measured and therefore replaced with the steering wheel angle and a lookup table which gives the function $\delta_{W12} = f(\delta_S)$ and considers the non-linear steering ratio.
Figure 2.7 shows that (5.3) is a purely geometric definition and valid if and only if:

1. $\alpha_{W12} = \alpha_{W34} = 0$, which means that the instantaneous center (IC) in Figure 2.7 moves to the right until IC is perpendicular beneath the rear wheel center of gravity;
2. the position of CG_v is known and therefore the distance between CG_v and the rear tires l_{34} is known.

Note that the two restrictions above are additionally needed to the assumptions for the single track model which we made in Section 2.5.1. Especially the requirement $\alpha_{W12} = \alpha_{W34} = 0$ is extremely restrictive and almost never fulfilled in practice. We can see from Figure 2.3b and (2.3b) that F_{yW} needs to be small to achieve cornering with $\alpha_{W12} = \alpha_{W34} \approx 0$, which is just the case during parking on small vehicle velocity.
Now the question should arise why is (5.3) worth to consider as path radius model although we know about the drastic restrictions? Why do we avoid to introduce more sophisticated path radius models which consider slip angles and tire models?
The motivation to use a rather simple path radius model is that contrary to Section 5.2.3, where the assumption $\theta = \theta_r$ was reasonable, it is unlikely that the path radius and road curvature radius coincidence. The reasons are that PSD usually does not cover:

1. the road curvature radius for each lane, hence the different path radius for each driving direction is not resolved;
2. disturbances through driver or traffic such as lane changes, passing and cutting of curves.

The first reason is a minor problem which could be resolved through road categories. The second reason is more serious and should emphasize that the prediction of lateral vehicle dynamics might introduce uncertainty in the prediction of the vehicle tractive force. Hence, it is more likely that the road curvature radius from PSD is only a rough approximation of the path radius and the simple path radius model in (5.3) is satisfactory.
Similar to Section 5.2.3, we apply the PKS from Section 3.9.1 with

$$\begin{aligned} B_t &= \sqrt{\left(\frac{l_{12}+l_{34}}{\tan \delta_{\mathrm{W}12,t}}\right)^2 + l_{34}^2} \\ \widehat{B}_t &\approx r_{,t}. \end{aligned}$$

Figure 5.3 shows the PKS estimate as path curvature (r^{-1}) and the PSD validation data for two independent test runs. Small curvature means straight driving, whereas large curvature indicates cornering. The correlation between PKS estimate and PSD validation data is worse than in Figure 5.2. However, if we keep in mind the discussion from above that predictive road curvature radius (r_{r}) data might only give a rough approximation for the actual path radius (and vice versa), PKS with (5.3) gives a good result. The general level of path radius variation is for PKS higher than for PSD. When PSD indicates straight driving ($r_{\mathrm{r}}^{-1} \approx 0$), PKS shows still some variation, which indicates that the driver performs also small steering wheel angle corrections to keep the lane during straight driving.
In specific, PKS follows PSD with good accuracy for $t = 100\,\mathrm{s}$ to $140\,\mathrm{s}$ in Figure 5.3a, which is a rather straight driving with minor curves. For $t = 140\,\mathrm{s}$ to $160\,\mathrm{s}$ PKS and PSD show a longer constant curve with a sharp

ending around $t = 160$ s. The approximately constant offset between PKS and PSD for $t = 140$ s to 155 s indicates the missing tire model in (5.3). During cornering, $\alpha_{W12} > 0$, which is not modeled in (5.3). Hence, the estimated path curvature is larger than the true (but herein unknown) path curvature. Another explanation for the constant offset between PKS and PSD is the discrepancy between center-line road curvature and current lane road curvature. PKS and PSD show good correlation during cornering for $t = 225$ s, 310 s, 405 s, and 540 s in Figure 5.3a. However, PKS indicates a sharp curve at $t = 510$ s which is not visible in PSD.

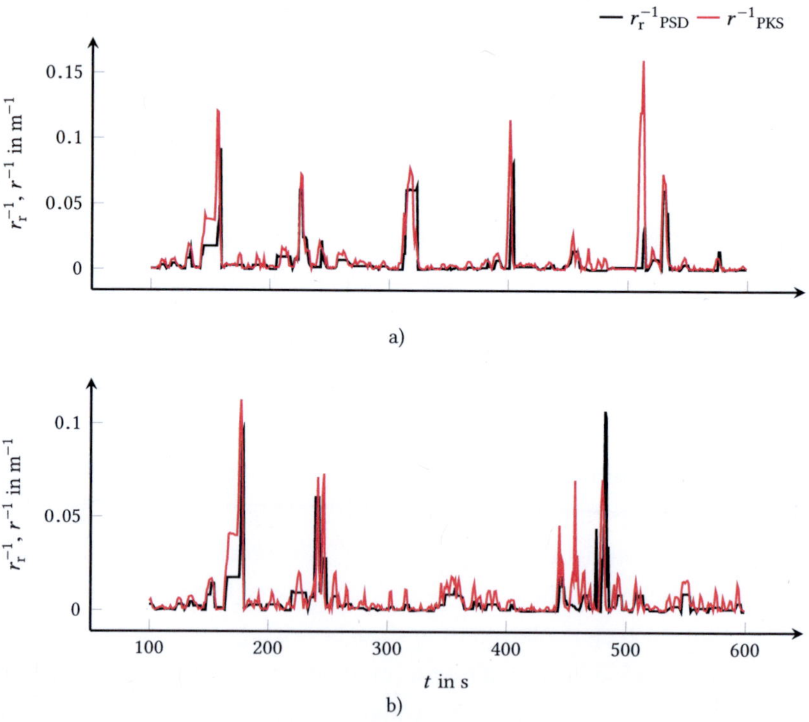

Figure 5.3: Path curvature estimation with PKS for two independent test runs. The road curvature validation data comes from predictive route data (PSD) [11, 119].

A disturbance through a lane change or passing might be the reason for this discrepancy between PKS and PSD.
Figure 5.3b shows all effects that we discussed for Figure 5.3a. Additionally, for $t \approx 490$ s PSD shows a sharp curve which is smoother estimated in the PKS result. A reasonable explanation is that the driver cut this sharp corner. To sum up, PKS and (5.3) give a reasonable approximation for the path radius with some uncertainty which we could also expect if we would use PSD road curvature radius data instead. However, ✎ the result interpretation suggests, that the dilemma between highly accurate path radius estimation on the one hand and uncertain road curvature radius predictive route data and disturbances caused by traffic and the driver on the other hand, requires more research.

5.3 Signal preprocessing

Halfmann and Holzmann [70, pp. 70–71] outline two principle methods for signal preprocessing in vehicle parameter estimation. The first is signal filtering. Most CAN signals are subject to noise. Especially the longitudinal acceleration sensor shows rich discretization error and noise in our data records. Halfmann and Holzmann [70, p. 70] list several signal filters (state filters, the Savitzky Golay filter, and adaptive filters) and give references where these filters were applied to vehicle mass estimation. The dilemma between low cutoff frequency and delay for low pass filters is emphasized. Most references of Table 4.1 use low pass filters.
Vahidi, Stefanopoulou, and Peng [181] used a second order butterworth filter with 25 Hz cut-off frequency to filter data which was recorded on 50 Hz. Bae, Ryu, and Gerdes [13] used a second order butterworth filter with lower cut-off frequency of 0.5 Hz on data which was read on 100 Hz. We used in [134, 135] a third order butterworth filter with 1 Hz cut-off frequency.

Figure 3.16 and the explanation in Section 3.9.1 lead us to use the PKS instead of low-pass filters herein because the delay of PKS is well defined and not frequency dependent. Moreover, we need derivatives from several CAN signals. The first derivative of the vehicle velocity is required in (5.2) and the polynomial-function approach gives smoother derivatives than the conventional used discrete derivative of low-pass filtered signals. Therefore, all CAN signals pass a bank of independent PKS with the same configuration. The delay of each smoothed CAN signal is exactly known and equal, which is adventitious compared with low-pass filters, where the delays of the filtered CAN signals depend on the frequency of each individual CAN signal.
The second signal preprocessing method which is given in [70, p. 71] is more trivial. Depending on the model structure, adverse driving styles which are not covered by the model should be neglected. Hence, a certain amount of data from the record is not used for parameter estimation. Neglecting adverse driving styles is to exclude outliers by a simple kind of outlier diagnostics. Trivial Boolean logic [134, 135] and [70, p. 72] or more advanced Fuzzy Logic [52] yield a trigger signal which activates the respective estimator. The trigger signal herein becomes true for the Boolean conditions:

1. no braking plus an additional half second after releasing the brakes;
2. the vehicle moves forward ($v_{\mathrm{V}} > 5.5\,\mathrm{m\,s^{-1}}$);
3. negligible slip inside the drive-train ($\dot{\theta}_{\mathrm{G}} \approx \dot{\theta}_{\mathrm{E}}$);
4. the absolute engine torque is above $10\,\mathrm{N\,m}$ ($|T_{\mathrm{E}}| > 10\,\mathrm{N\,m}$);
5. the drive-train model from Section 5.2.1 provides a rim torque signal.

The first condition excludes braking which is not modeled due to the expected strong variation of the braking friction coefficient while operation [157, p. 32] (see also the detailed explanation at the end of Section 2.7). The second condition ensures that the vehicle is in operation and the drive-train model operates well above quasi static friction. The third condition excludes shifting, which is not precisely modeled by the drive-train model.

The fourth condition ensures fully engaged tooth flanks in the gearbox and differential, and the fifth condition is trivial.

5.4 Vehicle tractive force models

As said in Section 1.1, the main focus of this work is to provide models which give a highly accurate prediction of the vehicle tractive force for look-ahead vehicle controllers. Hence, the model structure depends highly on available predictive data. A sole filtering problem such as vehicle side slip angle or vehicle center of gravity estimation allows more sophisticated model structures [79, 118] than a prediction problem, where we are restricted in the number of predictive data. For instance, the vehicle yaw rate $\dot{\psi}_{\mathrm{V}}$ plays a major role in vehicle side slip angle estimation. However, there is no predictive vehicle side slip angle data available for our problem. Hence, the following models have a rather simple model structure. We will not introduce sophisticated double lane non-linear vehicle dynamics models if we cannot feed them with predictive data in practice.
Moreover, we know from Section 2.4 that the predominant longitudinal forces and resistances which determine the vehicle tractive force depend on the vehicle velocity, its derivative, and second and fouth power. The acceleration force in (2.6) requires $\dot{v}_{\mathrm{V}}$, the longitudinal aerodynamic resistance in (2.7) v_{V}^2 (if we assume that $v_{\mathrm{a}} \approx v_{\mathrm{V}}$), and the coefficient of rolling resistance which determines the rolling resistance in (2.9) requires v_{V} and v_{V}^4. Add to this, the gradient angle needs to be known to model the climbing force in (2.5). Finally, we will consider the cornering resistance in (2.15) which is caused by lateral vehicle dynamics and requires the path radius and v_{V}^4 as measured input.
Hence, it makes intuitively sense to consider vehicle tractive force models which comprise:

1. the vehicle velocity from Section 5.2.2 as well as its derivative, second and fourth power,

2. the gradient angle from Section 5.2.3 and
3. the path radius from Section 5.2.4

in the measured inputs. The same signals serve later in Section 5.8 as validation data.

Why will we not use the road angle and road curvature radius from PSD as measured inputs, although it is apparent that we would require this predictive route data in vehicle look-ahead controllers? There are two reasons: First, only the most recent test runs from April 2014 contain PSD and we have seen in Figure 5.2 that the gradient angle gives a good approximation for the road angle from PSD and vice versa. On a lower extend, the same holds for the path radius and road curvature radius in Figure 5.3. Second, I cannot provide more detailed information about PSD. Hence, replacing the road angle and road curvature radius from PSD with clearly explained estimates of the gradient angle and the path radius makes this work more reproducible (see Section 5.2.3 and Section 5.2.4).

Now let us introduce a white-box model which represents the state of art in vehicle look-ahead controllers for the computation of the vehicle tractive force and two gray-box models which will be used with various recursive estimators of Chapter 3.

5.4.1 The abc longitudinal dynamics white-box model

White-box models with a_V, b_V, and c_V parameters (abc) are given in [24, 80, 110, 125] and the following model is similar and will serve as benchmark for the two subsequent gray-box models. The relation

$$\sum_{j_\mathrm{W}=1}^{4} \frac{T_{\mathrm{R},j_\mathrm{W}} - I_{\mathrm{W},j_\mathrm{W}} \ddot{\theta}_{\mathrm{W},j_\mathrm{W}}}{r_\mathrm{W}} = \frac{T_\mathrm{R} - I_\mathrm{W} \ddot{\theta}_\mathrm{W}}{r_\mathrm{W}} = F_{x_\mathrm{V}} = \dots$$

$$\dots = a_\mathrm{V} \cos\theta + b_\mathrm{V} \cos\theta v_\mathrm{V} + c_\mathrm{V} v_\mathrm{V}^2 + (g \sin\theta + \dot{v}_\mathrm{V}) m_\mathrm{V}, \quad (5.4)$$

considers vehicle longitudinal dynamics only and gives the abc white-box model which we will call $\mathcal{M}_0$ in the following. The subscript $_0$ denotes that all parameters of model $\mathcal{M}_0$ are assumed time-invariant. The a_V, b_V, c_V parameters come from vehicle coastdown experiments which were made by Porsche company.

The first line in (5.4) shows the transition from wheel individual forces into the single wheel model which is shown in Figure 2.8. The second line of (5.4) consists of four terms, whereas the first three terms involve the a_V, b_V, c_V parameters.

The first two terms in the second line of (5.4) give the rolling resistance of Section 2.4.4. The parameter a_V is a lumped parameter for $m_V g f_{r0}$, whereas $b_V = m_V g f_{r1}$. The relation between (5.4) and (2.9) is visible with $F_{zW} = m_V g \cos\theta$. Note that f_{r4} of (2.10) is neglected because we had no vehicle specific data for f_{r4} and f_{r4} is usually small, or in other words, often insignificant for the rolling resistance.

The third term in the second line of (5.4) is the longitudinal aerodynamic resistance which was introduced in Section 2.4.3. Here, c_V is a lumped parameter for $\frac{\rho_a}{2} A_V c_x$ and the connection to (2.7) is easily visible by assuming $\psi_a = 0$.

The last term in the second line of (5.4) is a shorthand notation of the climbing force (2.5) and acceleration force (2.6).

5.4.2 Longitudinal dynamics gray-box model

The first gray-box model in (5.5) is given in matrix notation and is called $\mathcal{M}_3$ in the following to indicate that any estimator must estimate three parameters. $\mathcal{M}_3$ is a purely longitudinal vehicle dynamics model and the relation to Section 2.4 is good visible if we compute the product AX of (5.5). The result is a sum of the rolling resistance (2.9), climbing force (2.5), acceleration force (2.6), and longitudinal aerodynamic resistance (2.7).

$$A = \begin{bmatrix} g\cos\theta & g\sin\theta + \dot{v}_{\mathrm{V}} & v_{\mathrm{V}}^2 \end{bmatrix} \tag{5.5a}$$

$$X = \begin{bmatrix} m_{\mathrm{V}} f_{\mathrm{r0}} & m_{\mathrm{V}} & \frac{\rho_{\mathrm{a}}}{2} A_{\mathrm{V}} c_x \end{bmatrix}^{\top} \tag{5.5b}$$

$$B = F_{x\mathrm{V}} = \sum_{j_{\mathrm{W}}=1}^{4} \frac{T_{\mathrm{R},j_{\mathrm{W}}} - I_{\mathrm{W},j_{\mathrm{W}}} \ddot{\theta}_{\mathrm{W},j_{\mathrm{W}}}}{r_{\mathrm{W}}} = \frac{T_{\mathrm{R}} - I_{\mathrm{W}} \ddot{\theta}_{\mathrm{W}}}{r_{\mathrm{W}}} \tag{5.5c}$$

We can apply $\mathcal{M}_3$ as multi-input-single-output output-error model of the form $AX \approx B$ (3.18) or as random-walk output-error model (3.48). Moreover, the multi-input-single-output errors-in-variables model (3.35) as well as the random-walk errors-in-variables model (3.50) give possible statistical assumptions to solve the unknown parameters of $\mathcal{M}_3$. All of these statistical models may be applied with regularized (or wind-up stable) and robust estimators.

5.4.3 Longitudinal and lateral dynamics gray-box model

The second gray-box model considers besides the vehicle longitudinal dynamics also lateral dynamics. $\mathcal{M}_4$ is shown in (5.6) and contains the additional fourth parameter $m_{\mathrm{V}}^2/c_{y\mathrm{W}}$ and a fourth measured input $v_{\mathrm{V}}^4/4r^2$ compared with $\mathcal{M}_3$ in (5.5). The product A_4X_4 of (5.6) gives the simple relation for the cornering resistance (2.15).

$$A = \begin{bmatrix} g\cos\theta & g\sin\theta + \dot{v}_{\mathrm{V}} & v_{\mathrm{V}}^2 & \frac{v_{\mathrm{V}}^4}{4r^2} \end{bmatrix} \tag{5.6a}$$

$$X = \begin{bmatrix} m_{\mathrm{V}} f_{\mathrm{r0}} & m_{\mathrm{V}} & \frac{\rho_{\mathrm{a}}}{2} A_{\mathrm{V}} c_x & \frac{m_{\mathrm{V}}^2}{c_{y\mathrm{W}}} \end{bmatrix}^{\top} \tag{5.6b}$$

$$B = F_{x\mathrm{V}} = \sum_{j_{\mathrm{W}}=1}^{4} \frac{T_{\mathrm{R},j_{\mathrm{W}}} - I_{\mathrm{W},j_{\mathrm{W}}} \ddot{\theta}_{\mathrm{W},j_{\mathrm{W}}}}{r_{\mathrm{W}}} = \frac{T_{\mathrm{R}} - I_{\mathrm{W}} \ddot{\theta}_{\mathrm{W}}}{r_{\mathrm{W}}} \tag{5.6c}$$

$\mathcal{M}_4$ is an extension of $\mathcal{M}_3$ and hence also suitable for the broad variety of estimators which were introduced in Chapter 3.

5.5 Parametrization of recursive estimators

This section shows the specific settings of several recursive estimators which solve the estimated parameters of $\mathcal{M}_3$ and $\mathcal{M}_4$. All recursive estimators were initialized with vehicle specific data for $\widehat{X}_{t-1}$ and the covariance matrix was initialized with $P_{t-1} = 100I$.

5.5.1 Recursive least squares

The RLS algorithm in Algorithm 3.1 requires only λ as user defined setup. The forgetting factor (λ) was set to 0.999 for $\mathcal{M}_3$ and $\mathcal{M}_4$.

5.5.2 Recursive regularized M-estimator

The settings of RRLM with robust estimated scale (Algorithm 3.4 in Line 6 of Algorithm 3.3) were: $\lambda = 0.999$, *type*–Levenberg-Marquardt regularization, influence function (ψ)–(3.15b), $\kappa = 1 \times 10^{-5}$, $\widehat{\sigma}_{t-1} = 25$, $\widehat{Q}_{t-1} = 350$, forgetting factor for $\widehat{\sigma}$ computation $\lambda = 0.995$, $\eta = 0.01$. These settings were similar for $\mathcal{M}_3$ and $\mathcal{M}_4$.

5.5.3 Stenlund-Gustafsson M-Kalman filter

The settings of SGMKF with robust estimated scale (Algorithm 3.16 in Line 6 of Algorithm 3.3) were: influence function (ψ)–(3.15b), $\widehat{\sigma}_{t-1} = 25$, $\widehat{Q}_{t-1} = 350$, forgetting factor for $\widehat{\sigma}$ computation $\lambda = 0.995$, $\eta = 0.01$ and $R^1 = 0.8$ for $\mathcal{M}_3$ and $\mathcal{M}_4$. However, P_d requires to distinct between $\mathcal{M}_3$ and $\mathcal{M}_4$. P_d was adjusted to

$$\mathrm{diag}(P_\mathrm{d}) = \begin{bmatrix} 5 \times 10^{-5} & 1 \times 10^{-6} & 1 \times 10^{-8} \end{bmatrix}^\top$$

for $\mathcal{M}_3$ and to

$$\mathrm{diag}(P_\mathrm{d}) = \begin{bmatrix} 5 \times 10^{-5} & 1 \times 10^{-6} & 1 \times 10^{-8} & 1 \times 10^{-6} \end{bmatrix}^\top$$

for $\mathcal{M}_4$, respectively. All off-diagonal entries of P_d were zero.

5.5.4 Recursive GTLS

The RGTLS estimator (Algorithm 3.8) was applied with $\lambda = 0.999$ and $\breve{P}_t = I$. Hence, RGTLS worked like an RTLS estimator.

5.5.5 Stenlund-Gustafsson IV M-Kalman filter

The parametrization of SGIVMKF with robust estimated scale (Algorithm 3.20 in Line 6 of Algorithm 3.11) was the same as Section 5.5.3 shows and the instruments were given by $\mathfrak{A}_t = A_{t-4}$.

5.6 Vehicle tractive force estimation

It makes sense to start the presentation and discussion of results with one recorded test run before we will explore the results for the prediction of the vehicle tractive force of all 81 test runs in Section 5.8. Moreover, we will focus within this section on $\mathcal{M}_3$, because the variety of models and estimators might confuse the reader. In sum, Section 5.8 will present results of 11 model-estimator combinations. Ten combinations arise from $\mathcal{M}_3$ and $\mathcal{M}_4$, which are estimated with RLS, RRLM, SGMKF, RGTLS and SGIVMKF, respectively. The eleventh model is $\mathcal{M}_0$, which does not require an estimator. Multiplied with the 81 test runs, the overall number of result sets is 891. Note that all results within this section ground on training data only. Model selection and model validation results in terms of performance indices that were discussed in Section 3.2 will be given by Section 5.8.
We will study now the properties of the four estimators: RLS, RRLM, SGMKF and RGTLS in terms of time-series plots for the estimated parameters in Figure 5.4. Remember Figure 3.1 for the correct classification of estimators and model structures. RLS is the recursive estimator for the multi-input-single-output output-error model (3.18) for instance.

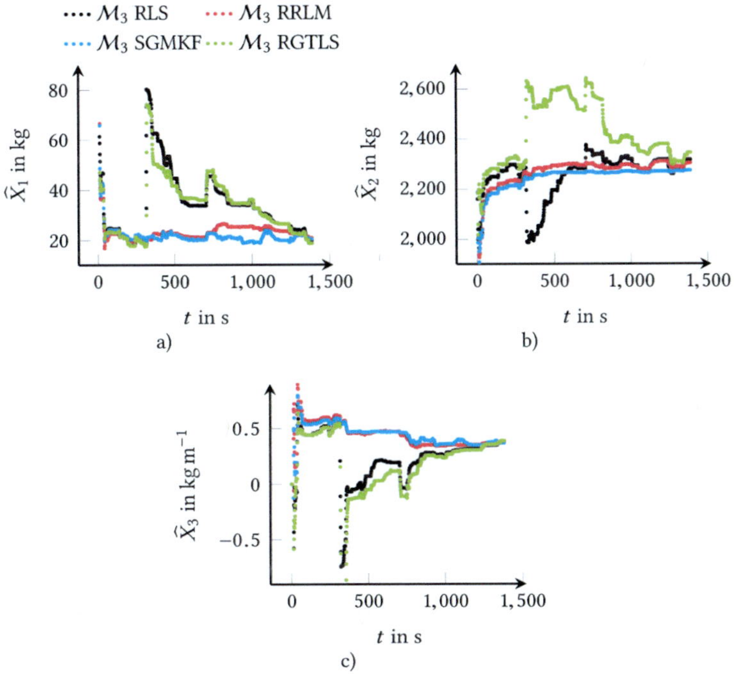

Figure 5.4: Estimated parameters for one test run $\mathcal{M}_3$ and various estimators.

The most notably observation in Figure 5.4 is the sudden change in all estimated parameters at $t = 400$ s for the RLS and RGTLS estimator. It is likely that there is an outlier in the data at $t = 400$ s. All estimated parameters have a physical meaning (and unit). Thereby, we know from experience that such drastic step changes in the parameters are unlikely or even impossible in practice. Moreover, negative values of $\widehat{X}_3$ in Figure 5.4c disagrees with the definition of the longitudinal aerodynamic resistance (2.7). The only practical explanation for $\widehat{X}_3$ being smaller than zero would be a strong wind gust from rear, because we assumed for $\mathcal{M}_3$ that $\psi_a = 0$.

However, such a strong wind gust is unlikely. Note that RLS and RGTLS converge after this step change at $t = 400$ s in the region of the robust and regularized or wind-up stable estimators RRLM and SGMKF. Hence, we can conclude that the recorded real world data requires robust estimators.

The results of RRLM and SGMKF are close. However, SGMKF shows a more stable solution for $\widehat{X}_2$ in Figure 5.4b, which presents the vehicle mass estimate. This better performance of SGMKF can be explained with the two different statistical models of RRLM and SGMKF. It was already said in Section 3.7 that the random-walk output-error model allows to consider an individual assumed variation rate for each parameter by adjusting Q or P_d, respectively. See also Figure 3.15 and Experiment 3.6 to get a similar observation for the errors-in-variables problem. However, we will stop here with the interpretation of parameter estimates and conclude that the estimated parameters of RRLM and SGMKF are in a reasonable range, with a small advantage for SGMKF. One reason to stop the evaluation of the parameter estimates is missing validation data for $\widehat{X}_1$ and $\widehat{X}_3$. Remember that no additional sensors were used in this work. However, as X_1 was defined as $m_\mathrm{V} f_{\mathrm{r}0}$ in (5.5), we can compute a rough estimate of $f_{\mathrm{r}0}$ for the SGMKF estimator from the converged results of Figure 5.4a and Figure 5.4b through $21/2250$, which gives $f_{\mathrm{r}0} = 0.0093$. This result is reasonable for modern tires which are optimized for reduced coefficient of rolling resistance, where $f_\mathrm{r} \approx 1\,\%$ [179, Figure 3.4/2]. We will discuss vehicle mass estimation results in the forthcoming Section 5.9 in more detail.

The final part of this section is a result discussion of output corrections for RLS, RRLM, SGMKF and RGTLS with Figure 5.5, Figure 5.6, and Figure 5.7. All Figures show the same data, but in a different representation. Figure 5.5 gives the output corrections as time-series plots for all four estimators. The likely outlier at $t = 400$ s is well visible through the sudden large output correction in Figure 5.5a–Figure 5.5d. Additionally, a second possible outlier is discovered through Figure 5.5 at $t = 750$ s. Note that the output corrections of RGTLS in Figure 5.5d at $t = 400$ s and 750 s are the smallest of

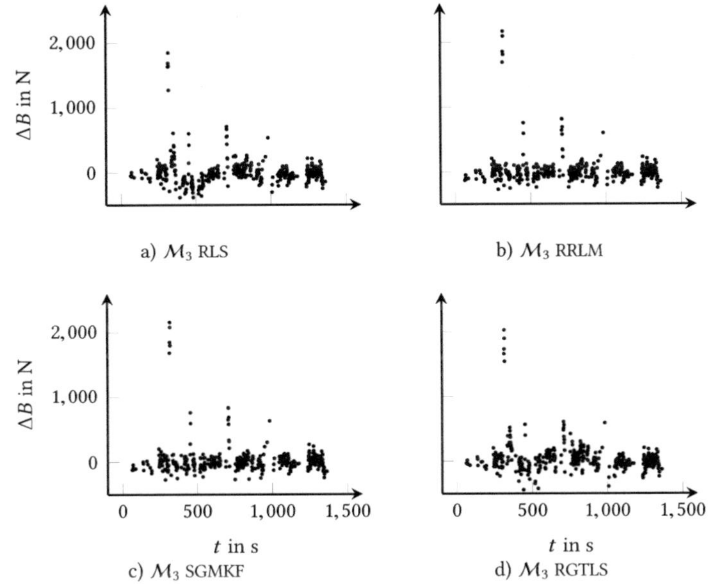

Figure 5.5: Output corrections for one test run and various estimators.

all four estimators, followed by RLS in Figure 5.5a. The output corrections of RRLM and SGMKF at t = 400 s and 750 s in Figure 5.5b and Figure 5.5c, respectively, are close and both estimators show larger output corrections compared with RGTLS and RLS. In conclusion, robust estimators such as RRLM and SGMKF show larger output corrections when outliers are present than non-robust estimators such as RLS and RGTLS.

The reason for the larger output correction of robust methods is to find in the cost functions. If we compare the cost function of RLS (3.24) with the cost function of RLM in (3.25) (let us skip the regularization for a moment, then RRLM becomes RLM) and focus on the part where the output correction (written as $B_i - A_i X$) is involved, we can see that the non-robust RLS estimator gives the smallest possible output correction in the least-squared sense. This observation is important to decide for an appropriate performance index in Section 5.8. The estimator with the smallest output

correction is not the best estimator for our application, or as we can see from Figure 5.4, the estimator with the smallest output correction (recursive GTLS) is even the worst estimator.
Figure 5.5 indicates a second observation. The non-robust estimators RLS and RGTLS show a larger dispersion of the output correction than the robust estimators RRLM and SGMKF. Compare the trend of the output corrections for RLS in Figure 5.5a for $t = 400$ s to 750 s with RRLM in Figure 5.5b for the same period. RLS is quite long disrupted and requires a lot of new data until the output corrections disperse once again symmetrically around zero, whereas RRLM does not show a distinctive disruption. However, this observation is better visible in Figure 5.6.

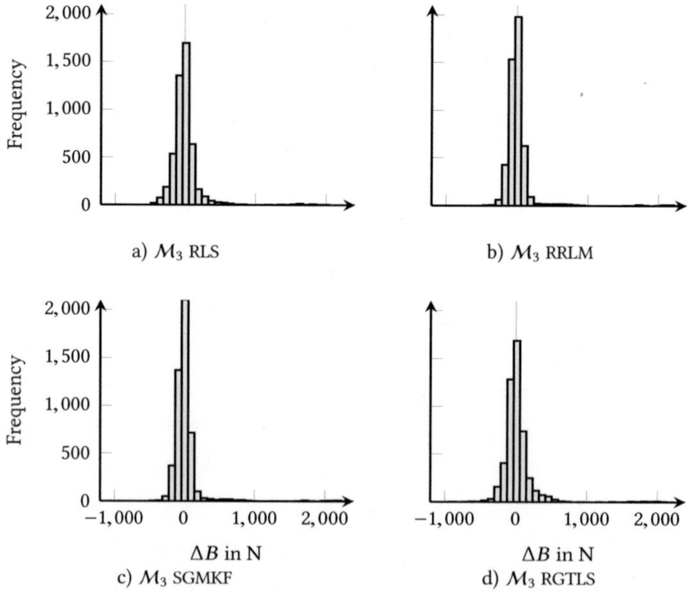

Figure 5.6: Histograms of the output correction for one test run and various estimators.

Figure 5.6 presents the output corrections of all four estimators as histogram. All histograms show heavy-tailed distributions, that indicate small fractions of large output corrections, which are quite likely outliers. The output correction dispersion of robust estimators (RRLM in Figure 5.6b and SGMKF in Figure 5.6c) is smaller than the output correction dispersion of the non-robust estimators (RLS in Figure 5.6a and RGTLS in Figure 5.6d). Although on small frequency, there are some output corrections in the range of 2000 N visible in Figure 5.6. These drastic outliers are responsible for the breakdown of RLS and RGTLS. The M-estimators RRLM and SGMKF are superior robust in this case because the outliers occur in the output corrections. It appears that there are no bad leverage points present in the data, remember the explanation about breakdown point, leverage point, and M-estimators in Section 3.4.2.

Figure 5.7 presents scatter plots of the estimated output over the measured output for all four estimators. The diagonal thin line is the ideal relation between estimated output and measured output. We can see both observations from above. First the small distribution for the majority of scatter points of the robust estimators RRLM and SGMKF in Figure 5.7b and Figure 5.7c, respectively. The second observation, which was the larger output corrections of RRLM and SGMKF compared with RLS and RGTLS, is visible if we take into account that the output correction is the vertical distance between estimated output and measured output. Hence, the far outlying points in Figure 5.7b and Figure 5.7c represent the large output correction of the robust estimators.

We can use Figure 5.7 to classify the scatter points into two groups: majority and minority. A way to sketch a borderline between these groups could be an ellipse with a major axis which is collinear with the diagonal thin line in Figure 5.7. The perpendicular minor axis of the ellipse would give the dispersion of the estimators. We would probably pick RRLM or SGMKF once again from this procedure in Figure 5.7, because the length of the minor axis of the described ellipse is smaller compared with the non-robust estimators

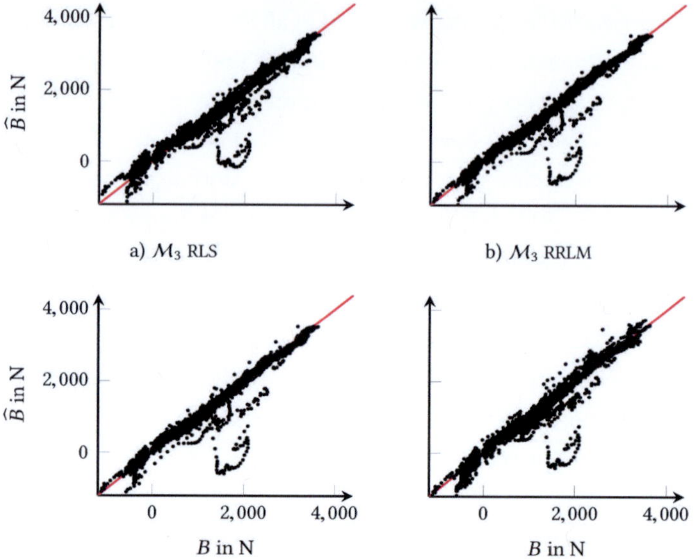

a) $\mathcal{M}_3$ RLS

b) $\mathcal{M}_3$ RRLM

c) $\mathcal{M}_3$ SGMKF

d) $\mathcal{M}_3$ RGTLS

Figure 5.7: Scatter plots for one test run and various estimators. Outliers are well visible and sophisticated outlier diagnostics ([143]) could be used to identify adverse driving styles.

RLS and RGTLS. Hence, Figure 5.7 is probably the best way to discover outliers within Figure 5.5–Figure 5.7 through a plot. However, we will not analyze any further specific situations (experimental conditions) where outliers probably occurred. The robust estimators give superior results and it is not ensured that even if we solved one specific issue, for instance if we adjust the values of the Boolean logic in Section 5.3, that this treatment would work for all test runs. Hence, we have to accept that there are outliers in the recorded data and apply preferably robust estimators.

The discussion upon here did not emphasize the contribution of regularization or wind-up avoidance to ensure stable results of RRLM and SGMKF, respectively. However, the key to reliable and stable estimates from this recorded data is a combination of robustness and regularized or wind-up stable estimators. Remember the third challenge, poor excitation, which was mentioned in Section 5.1 and observed from Figure 5.1.

We would need to vary the regularization parameter between zero (no regularization) and some upper limit in numerous subsequent simulations on the same data to measure the contribution of regularization which lead to the stable RRLM estimates. This procedure would give us an optimal regularization parameter and is known under the term L-curve, which is well explained in [160, Figure 1.1] and [95, pp. 52–57] (see also Section 3.3). The L-curve is a batch method and because of this not ideal for our needs. However, Van Waterschoot, Rombouts, and Moonen [186] provide recursive methods that give an optimal regularization parameter in a acoustic signal enhancement application.

I do not claim that the settings for RRLM made in Section 5.5.2 are optimal in a mathematical sense. The same holds for the other estimators and this discussion refers us back to Section 3.10.2 where we talked about IAE and MME, which are basically methods to tune estimators automatically.

Therefore, ✎ let us add the question: *how to chose an optimal regularization parameter recursively for vehicle tractive force estimation* to the list of open research questions within this dissertation.

However, another way is possible to get a rough impression how much regularization we need for this application. Let us compare the vehicle mass estimates of RLS and RGTLS in Figure 5.4b. Although neither RLS nor RGTLS involve regularization, we can use these estimators to compare what happens if we add deregularization to an estimator. Remember that RGTLS is a TLS-based estimator and TLS lacks under deregularization. This was discussed in the open topics of Section 3.6.3. If we look closer what happens after the first likely outlier within the period $t = 400$ s to 750 s in Figure 5.4b, we can observe that the RLS estimate converges towards the reliable vehicle mass estimates of RRLM and SGMKF, whereas RGTLS does not. Moreover, the RGTLS estimate shows a sine wave like curve. In short, the deregularized RGTLS estimator becomes uncertain. If we compare this observation with the period after the second likely outlier for $t > 750$ s, we can see that RLS and RGTLS converge to the RRLM and SGMKF estimate. These two different manners of RGTLS, after the first likely outlier not converging and after the second likely outlier converging, indicates that there is no sufficient rich excitation in the period $t = 400$ s to 750 s in order to feed RGTLS with enough useful new information. Hence, carefully applied regularization helps to ensure stable estimates during periods of poor excitation. See also Figure 3.8 and Experiment 3.3 where RLM and RRLM were compared.

5.7 Robust performance index

Remember that the discussion of results within Section 5.6 was only made for one test run and one model. It is impossible (or would require several hundred pages) to proceed the interpretation of results in the same way (figures for the estimated parameters and output corrections) for all 891 tuples of test runs, models, and estimators. Hence, we seek for a reliable scalar measure that gives a ranking for all model estimator combinations for each test run. Such performance indices were shown in Section 3.2.1 and the mean squared error in (3.3a) is probably the most prominent performance index from the list in (3.3). However, MSE itself is not robust.

What does MSE mean? The mean squared error involves the arithmetic mean ($\mu(\cdot)$), which is a non-robust location estimator. We could replace $\mu(\cdot)$ with an M-estimator [81], but the median ($\mathrm{med}(\cdot)$) is even more robust and simple to compute. Hence, we will use the median squared error (MEDSE)

$$\mathrm{MEDSE} = \underset{i}{\mathrm{med}}\left(\left(B_i - \widehat{B}_i\right)^2\right) \tag{5.7}$$

as robust performance index in Section 5.8. We can easily justify this decision if we take a short look on Figure 5.5 and imagine that we would plot the squared output correction instead of the shown output correction. The arithmetic mean would fail as location estimator in this case, because the large output corrections would be even larger compared with the majority of output corrections because of the squares. The median instead gives us a robust estimate of the location. Zoubir et al. [200, Figure 1] give an excellent example in which the arithmetic mean, trimmed arithmetic mean, and median are compared in a location estimation problem in the presence of outliers. The breakdown point of the arithmetic mean is 0 %, whereas the breakdown point of the median is 50 %. Moreover, [200, Figure 1] shows that the arithmetic mean gives a location estimate which is aligned towards the outlier away from the majority of data. As RRLM and SGMKF produce the largest output corrections, the MSE for these estimators would be bigger then the MSE for RLS and RGTLS. Hence, a MSE-based performance index might lead us to wrong conclusions.

5.8 Vehicle tractive force prediction

This section answers the question which model and estimator combination gives the most accurate prediction of the vehicle tractive force. We will begin with $\mathcal{M}_3$ on one recorded test run and then switch to the full range of models and estimators applied on all 81 test runs.

Cross-validation is the most powerful model selection method and gives in parallel a result for model validation, see Section 3.2. In order to apply cross-validation, each test run was divided into training data and validation data, where the validation data is a block of samples at the end of each record. The size of validation data varies in the number of prediction steps with $k = 300, 900, 1200, 1500$, and 1800, and only samples with a true trigger signal were used. Remember the Boolean logic which provides a trigger signal to exclude adverse driving styles. The recursive estimators work on 10 Hz. Hence, the validation data resembles a minimum prediction time of 30 s, 90 s, 120 s, 150 s, and 180 s. However, a trigger signal with 300 true samples may require more than 30 s of data if the trigger signal shows some false samples.

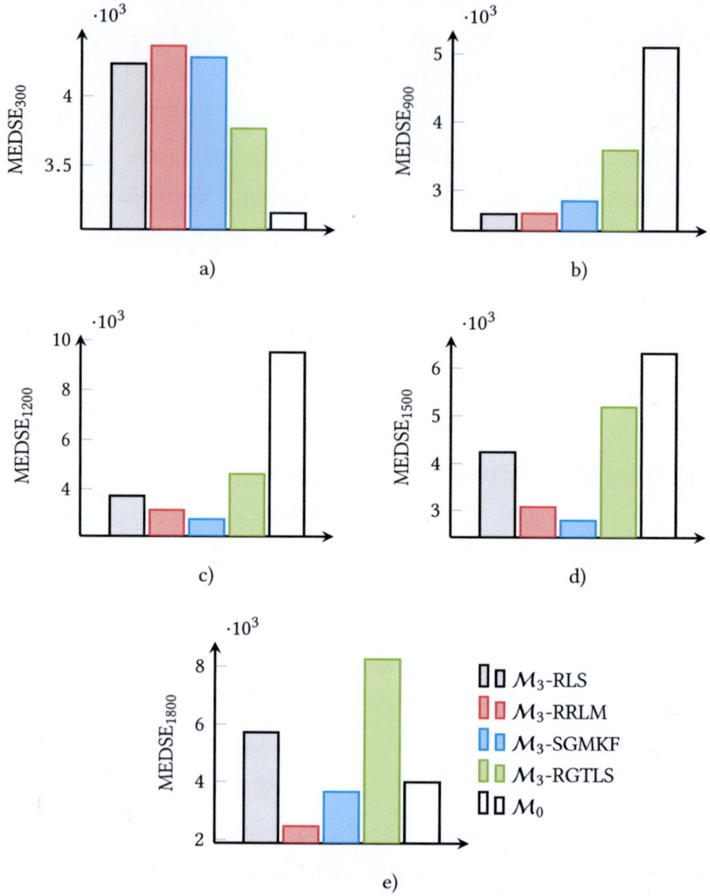

Figure 5.8: Model selection in terms of the vehicle tractive force prediction, measured by the MEDSE performance index for one test run and two models. $\mathcal{M}_0$ is a white-box model whereas $\mathcal{M}_3$ is a gray-box model which was identified with four recursive estimators. The number of prediction steps in the validation data is $k = 300$, 900, 1200, 1500, and 1800, which is denoted by the subscripts at MEDSE, respectively. This subscript corresponds to a prediction time of at least 30 s, 90 s, 120 s, 150 s, and 180 s.

Figure 5.8 gives the prediction MEDSE (5.7) of the vehicle tractive force for one test run of the white-box model $\mathcal{M}_0$ and the gray-box model $\mathcal{M}_3$ with the estimators RLS, RRLM, SGMKF and RGTLS.
Figure 5.8a shows the result for the smallest validation data with $k = 300$. Here, $\mathcal{M}_0$ is the best, followed by $\mathcal{M}_3$-RGTLS, $\mathcal{M}_3$-RLS, $\mathcal{M}_3$-SGMKF, and $\mathcal{M}_3$-RRLM. This superior performance of $\mathcal{M}_0$ will be a rare case in the following.
Figure 5.8b–Figure 5.8e show that there is always a better gray-box model in the set of candidate models available than $\mathcal{M}_0$. Moreover, the prediction quality of $\mathcal{M}_3$-RGTLS and $\mathcal{M}_3$-RLS is unstable. $\mathcal{M}_3$-RGTLS and $\mathcal{M}_3$-RLS work well in Figure 5.8b and Figure 5.8c, suffer in Figure 5.8d, and fail drastically in Figure 5.8e compared with $\mathcal{M}_3$-RRLM and $\mathcal{M}_3$-SGMKF. $\mathcal{M}_3$-RRLM gives the best overall accuracy, followed by $\mathcal{M}_3$-SGMKF, which is inferior in Figure 5.8e. Yet, it has the largest prediction horizon and gives thereby the most challenging validation data.
Apart from Figure 5.8a, $\mathcal{M}_0$ is clearly the worst choice compared with $\mathcal{M}_3$-RRLM and $\mathcal{M}_3$-SGMKF. The inferior prediction quality of $\mathcal{M}_0$ is an important observation for vehicle look-ahead controllers which seek for energy efficient driving, because all of the references that were given in Section 1.1 use models that are similar to $\mathcal{M}_0$. It is likely that the mentioned vehicle look-ahead controllers are influenced in their actions by the prediction quality of vehicle tractive force models. Therefore, ✎ the application of vehicle look-ahead controllers in conjunction with adaptive gray-box models from this dissertation might offer much research potential.
On the other hand, Figure 5.8 is just a snapshot of a single test run. So, we can ask: is $\mathcal{M}_0$ just by accident inferior in Figure 5.8b–Figure 5.8e?
First we need a compact representation of the performance index (prediction MEDSE of the vehicle tractive force) which allows to select a model over all test runs. A boxplot is the ideal plot for this purpose. A boxplot contains information about the location and dispersion of data. Moreover, boxplots allow to decide if the location between two classes of data differ significantly.

A boxplot shows the median, which is the 0.5 quantile, as central line inside a box. The lower and upper edge of the box depict the 0.25 and 0.75 quantile, respectively. Two antennas arise below and above the box and the end of each antenna is marked with a whisker. The lower and upper whisker are computed by $Q_{0.25} - 1.5(Q_{0.75} - Q_{0.25})$ and $Q_{0.75} + 1.5(Q_{0.75} - Q_{0.25})$, respectively [58]. Data outside of the whiskers denote outliers. Two medians differ significantly at 5 % significance level if their significance intervals do not overlap. McGill, Tukey, and Larsen [111] give the lower and upper significance level of the median with $Q_{0.50} - 1.57(Q_{0.75} - Q_{0.25})/\sqrt{m}$ and $Q_{0.50} + 1.57(Q_{0.75} - Q_{0.25})/\sqrt{m}$, respectively.
The idea is now to show the performance index for each test run as single dimension boxplot for the white-box model $\mathcal{M}_0$ and each combination of gray-box models, $\mathcal{M}_3$, $\mathcal{M}_4$ with estimators RLS, RRLM, SGMKF, RGTLS, SGIVMKF. Note that SGIVMKF appears within this chapter for the first time. SGIVMKF was simply not needed for the result interpretation upon here.
Figure 5.9 shows the prediction quality as MEDSE of the vehicle tractive force as boxplot of all models and estimators for 60 test runs. The number of usable data records is reduced from 81 to 60 because $k = 900$ requires rather long test runs. Twenty one test runs were simply too short to split the data appropriately into training data and validation data. Note that the ordinate of Figure 5.9 is logarithmic, which is uncommon for boxplots. Hence, the significance levels of $\mathcal{M}_0$ and to a lower extend of $\mathcal{M}_3$ and $\mathcal{M}_4$ appear asymmetric but in fact they are symmetric. The reason to use a logarithmic ordinate is to put all models into one diagram and to resolve the lower values more precisely.
The most striking observation is that the median of $\mathcal{M}_0$ ranges significantly above all gray-box models. Hence, the prediction quality of $\mathcal{M}_0$ is significantly inferior to any gray-box model. We can state this significant difference by comparing the lower significance level of $\mathcal{M}_0$ (the line below

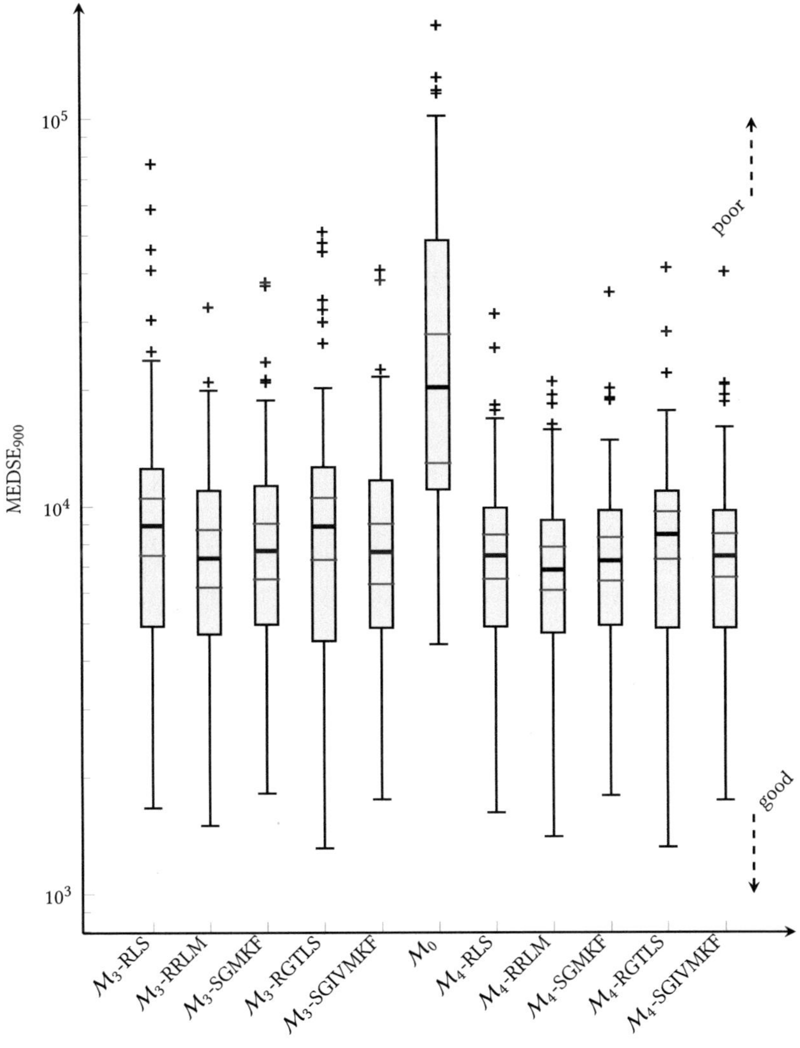

Figure 5.9: Boxplot of the prediction quality in terms of the vehicle tractive force measured with MEDSE. The number of prediction steps is $k = 900$ and the number of usable test runs is $m = 60$.

the central median line) with all other upper significance levels, because there is no overlap in the significance levels.
However, there are some boxes which overlap with $\mathcal{M}_0$ as well as all antennas, whiskers and outliers. In other words, there are rare test runs where $\mathcal{M}_0$ was as good as any of the gray-box models or probably better.
Why gives $\mathcal{M}_0$ sometimes satisfactory results? If we go back to (5.4) and its explanation, we see that the parameters (a_V, b_V, c_V, m_V) of $\mathcal{M}_0$ come from coastdown measurements. Accordingly, $\mathcal{M}_0$ gives a good prediction quality if the experimental condition in the test runs meets roughly the experimental condition from the coastdown experiment. If the number of passengers, air temperature, air pressure, tire-pressure, tire-temperature, gear-box oil-temperature, and many more experimental conditions correspond to the conditions during the coastdown test, $\mathcal{M}_0$ performs well and gray-box models are not needed.
On the other hand, Figure 5.9 tells that the experimental conditions that lead to $\mathcal{M}_0$ during the coastdown experiment and during the performed test runs differ a lot. $\mathcal{M}_0$ assumes that the parameters are time-invariant, which is a typical characteristic of white-box models. However, the consideration of time-varying parameters in $\mathcal{M}_3$ and $\mathcal{M}_4$ is clearly superior for the majority of test runs. Moreover, Figure 5.9 shows even on logarithmic scale that the innerquartile range (the height of the box) of $\mathcal{M}_0$ is larger than the innerquartile range of any gray-box model. Consider that the innerquartile range of $\mathcal{M}_3$ and $\mathcal{M}_4$ appears magnified due to the logarithmic ordinate compared with $\mathcal{M}_0$. Accordingly, the innerquartile range gives an additional argument for the conclusion that $\mathcal{M}_3$ and $\mathcal{M}_4$ in combination with any of the presented estimators give a clearly improved prediction quality.
A deeper look at Figure 5.9 shows that there is a second important but not significant difference in the medians to find. The lower significance level of $\mathcal{M}_3$-RLS overlaps only in a small extend with the upper significance level of $\mathcal{M}_4$-RRLM. In other words $\mathcal{M}_4$-RRLM gives a better but not significantly better prediction quality in terms of the median than $\mathcal{M}_3$-RLS. Moreover,

$\mathcal{M}_4$-RRLM shows the lowest median of all models and is the best choice from a model selection and a model validation point of view. Additionally, the innerquartile range of $\mathcal{M}_4$-RRLM is the smallest in Figure 5.9, which is another indicator that $\mathcal{M}_4$-RRLM is the best choice for a high prediction quality of the vehicle tractive force.

If we compare the results of $\mathcal{M}_3$ and the respective estimators the ranking from worst to best in terms of the median becomes: $\mathcal{M}_3$-RLS, $\mathcal{M}_3$-RGTLS, $\mathcal{M}_3$-SGMKF, $\mathcal{M}_3$-SGIVMKF, and $\mathcal{M}_3$-RRLM. Accordingly, the non-robust and non-regularized estimators are the worst, followed by two robust Kalman filters (with regard to which the instrumental variables Kalman filter shows the better result) and the robust and regularized RRLM estimator. Although the medians do not differ significantly, RRLM is the best choice for $\mathcal{M}_3$.

The ranking of estimators from worst to best for $\mathcal{M}_4$ differ to the ranking for $\mathcal{M}_3$ and becomes: $\mathcal{M}_4$-RGTLS, $\mathcal{M}_4$-SGIVMKF together with $\mathcal{M}_4$-RLS, $\mathcal{M}_4$-SGMKF, and $\mathcal{M}_4$-RRLM. Hence, the ranking in $\mathcal{M}_4$ can not be split into non-robust and robust estimators, because $\mathcal{M}_4$-SGIVMKF is slightly worse than $\mathcal{M}_4$-RLS. On the other hand, the main result from $\mathcal{M}_3$ holds also for $\mathcal{M}_4$, namely that $\mathcal{M}_4$-RRLM is the best estimator inside $\mathcal{M}_4$.

To sum up, $\mathcal{M}_4$-RRLM gives better prediction quality than $\mathcal{M}_3$-RLS and $\mathcal{M}_4$-RRLM, and is significantly better than $\mathcal{M}_0$. Moreover, $\mathcal{M}_4$-RRLM gives the lowest median and the smallest innerquartile range. Therefore, $\mathcal{M}_4$-RRLM is the superior model estimator combination for vehicle tractive force prediction.

5.9 Vehicle mass estimation

Table 4.1 shows that highly precise vehicle mass estimation is a topic of great importance in research. Although vehicle mass estimation is not the primary goal of this work, we will compare the results of $\mathcal{M}_3$ and $\mathcal{M}_4$ in conjunction with RLS, RRLM, SGMKF, RGTLS, and SGIVMKF in terms of the performance index NMSE which is given in (3.3b). It is save to use the non-robust performance index NMSE here, because the validation data is simply a scalar which comes from balancing the vehicle and passengers prior driving. Hence, the validation data is free of outliers. I balanced each vehicle once and added the weight of passengers, luggage, and fuel in 44 test runs. Hence, the number of test runs with vehicle mass validation data is 44. Note that a small but negligible error arises due to fuel consumption that is not considered if we balance the vehicle only once at the beginning of each test run.

Furthermore, there is no need to use cross-validation, which requires to split the data into training data and validation data, to measure the accuracy of vehicle mass estimates. However, it makes intuitive sense to measure the accuracy of vehicle mass estimates not directly from start of each test run. It is useful to consider a certain period which is required by each recursive estimator to converge. Remember that each estimator must be initialized and it takes some iterations until the effect of initialization diminishes. One way could be to define a minimum number of iterations which have to be performed by each estimator before its estimates can be used. However, during poor excitation the number of iterations does not say something about the amount of information. Actually, the correlation between number of iterations and amount of information depends much on data. Therefore, another criteria is required to decide if an estimator gives converged estimates or not.

Herein, an upper bound of the covariance matrix was defined. Remember that the covariance matrix was initialized with $P_{t-1} = 100I$ in Section 5.5. The convergence test is simply done by Boolean logic with $P_t < P_{\max}$. The convergence of SGMKF was checked and if this test was true, the results of all estimators were assumed to be converged. The choice of SGMKF is of course arbitrary, but makes sense, because it is a robust estimator which has a lower convergence rate than the two non-robust estimators RLS and RGTLS, compare Section 3.4.1. Actually, RGTLS in Algorithm 3.8 does not provide information about the parameter uncertainty. This problem was only recently solved in [136]. Hence, it was not possible to use RGTLS as indicator for convergence.

Figure 5.10 gives the vehicle mass NMSE for all gray-box models, where the best possible NMSE is one. Hence, the best gray-box model is the model with the box which is closest to one. RGTLS provides poor results in both models ($\mathcal{M}_3$ and $\mathcal{M}_4$), and $\mathcal{M}_3$-RGTLS is significantly poorer than any other estimator. However, the significance intervals of $\mathcal{M}_3$-RLS, $\mathcal{M}_4$-RLS, and $\mathcal{M}_4$-RGTLS overlap, which means that $\mathcal{M}_4$-RGTLS is not significantly poorer than RLS. As said, RGTLS suffers from missing robustness and regularization when applied to the recorded data. However, RGTLS performs well if the data does not contain challenges like outliers and poor excitation as reported in [133, 136] and reported from a related vehicle mass application where RGTLS was recently applied [6].

If we skip RGTLS and focus on the median, the ranking from worst to best becomes: $\mathcal{M}_4$-RLS, $\mathcal{M}_3$-RLS, $\mathcal{M}_3$-RRLM, $\mathcal{M}_4$-RRLM, $\mathcal{M}_3$-SGMKF, $\mathcal{M}_4$-SGMKF, $\mathcal{M}_3$-SGIVMKF, and $\mathcal{M}_4$-SGIVMKF. Note that all significance intervals overlap. Hence, there is no significant best model estimator combination. The best choice is $\mathcal{M}_4$-SGIVMKF. This gray-box model gives the highest vehicle mass NMSE and a small innerquartile range. However, the silbling $\mathcal{M}_3$-SGIVMKF shows a smaller innerquartile range. The inclusion of prior knowledge about time-varying parameters on different rate through the random-walk model and the use of instruments improve the accuracy in vehicle mass estimation.

$\mathcal{M}_4$ gives in vehicle tractive force prediction (Section 5.8) and in vehicle mass estimation better results than $\mathcal{M}_3$. However, the estimator of choice differs in prediction or estimation. Section 5.8 dealt with a prediction problem and $\mathcal{M}_4$-RRLM was the best model estimator combination. This section however deals with a parameter estimation problem and the random-walk errors-in-variables model, which is solved by SGIVMKF, outperforms the multi-input-single-output output-error model, which is solved by RRLM.

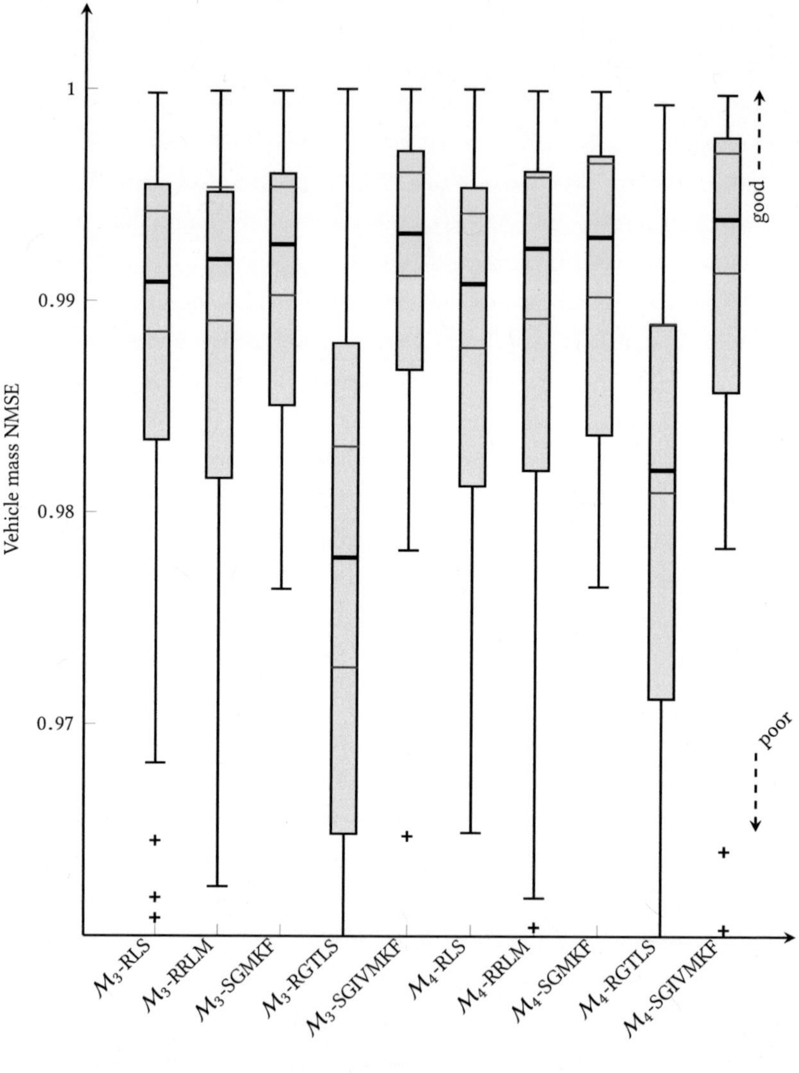

Figure 5.10: Normalized mean squared error of vehicle mass estimates for various models and estimators. The number of usable test runs was $m = 44$.

This result matches nicely with the recommendation of the first question in the hands on guideline for choosing the most appropriate estimator in Section 3.11, where it was recommended to favor the random-walk errors-in-variables model in parameter estimation. In conclusion, $\mathcal{M}_4$-SGIVMKF is the best choice for vehicle mass estimation.

NMSE can be converted into a relative error with $\sqrt{1 - \text{NMSE}}$. Hence, the relative vehicle mass error of $\mathcal{M}_4$-SGIVMKF becomes 7.75 %, with NMSE roughly 0.994, which is the median of $\mathcal{M}_4$-SGIVMKF. This result is rather poor for vehicle mass estimators compared with related work in Table 4.1 and own work [134, 135] which shows vehicle mass estimators which converge within an relative error of ±1 %.

However, consider that the recorded data is more challenging due to poor excitation and outliers. Moreover, the results herein in terms of highly accurate vehicle mass estimates may be easily improved by adjusting the values of the Boolean logic in Section 5.3. One action could be to improve the signal to noise ratio by enhancing the minimum engine torque. In addition to this, the drive-train model is of rather simple model structure. For instance, the wheel moments of inertia were not adjusted between summer and winter tires and a biased wheel moment of inertia causes the rim torque to be biased, which explains the bias in the vehicle mass estimates. However, the major reason for the rather poor vehicle mass estimates is that the drive-train model shares the same model structure for the two wheel drive and the four wheel drive vehicle. Hence, some internal losses in the four wheel drive vehicle were not sufficiently modeled. Hence, the rim torque must be biased in the records of the four wheel vehicle.

The ✎ intention of Figure 5.10 is to provide a comparison of various estimators for vehicle mass estimation. The random-walk errors-in-variables model, which is represented by SGIVMKF, has not been used in literature before, but offers potential to improve the accuracy of related vehicle mass estimators.

Summary

This chapter studied the prediction quality of a white-box model, two gray-box models, and various estimators from Chapter 3 with respect to accurate prediction of the vehicle tractive force. The model selection grounds on numerous test runs which were conducted on public roads under varying environmental conditions. The best combination of model and estimator is $\mathcal{M}_4$-RRLM, which is a linear gray-box model which considers vehicle longitudinal and lateral dynamics. The RRLM estimator is a recursive regularized estimator for the linear multi-input-single-output output-error model. Vehicle mass estimation results were discussed to a smaller extend. The best model which gives the most accurate vehicle mass estimates is $\mathcal{M}_4$-SGIVMKF. The SGIVMKF estimator denotes a robust and wind-up stable Kalman filter with instruments and solves the linear random-walk errors-in-variables model.

6 Concluding remarks

"All of science is nothing more than the refinement of everyday thinking."

Einstein [48, p. 313].

Outline

This chapter gives a summary of conclusions which were made throughout this dissertation and depicts the main contributions of this work. However, the emphasis of this chapter is a summary of open problems which are worth to consider for further research.

6.1 Conclusions

The conclusions started in Chapter 4, which gives a literature review of related research.

To the best of my knowledge, there is no direct related research which has been conducted with the aim to built models which give a precise prediction of the vehicle tractive force. The literature search discovered mostly vehicle parameter estimators with a strong focus on vehicle mass estimation.

Recursive least squares (RLS) was the prior estimator, followed by Kalman filters. The predominant model structure was a linear vehicle longitudinal dynamics gray-box model. There are attempts considering parameters which vary on different rate, such as recursive least squares with multiple forgetting (RLSmf). However, the connection between RLS and the Kalman filter has not been reported or used in related literature.

The need of rich excitation was several times emphasized in related research. However, rich excitation cannot be ensured in real world applications because rich excitation contradicts important aims like high driving comfort and high energy efficiency of vehicles. Poor excitation is typically treated by exclusion of data. However, the connection between poor excitation, ill-posed problems, and wind-up has not been reported in the related literature. Moreover, methods which stabilize recursive estimators such as regularization or wind-up stable filters have not been considered. Hence, the scope of estimators, which were used in vehicle parameter estimation, was limited. This was the motivation to write the detailed survey of models and estimators in Chapter 3.

The predominant conclusion of Chapter 5 was that all presented gray-box models gave significant better prediction quality for the vehicle tractive force than the white-box model with a_V, b_V, and c_V parameters (abc), which is state of the art in vehicle look-ahead controllers seeking energy efficient driving (see the references in Section 1.1).

The model offering the best prediction of the vehicle tractive force from the set of candidate models is model $\mathcal{M}_4$ in conjunction with the recursive regularized M-estimator (RRLM). $\mathcal{M}_4$-RRLM stands for a vehicle longitudinal and lateral dynamics gray-box model with a robust, regularized, recursive estimator, which solves the multi-input-single-output output-error model. Assuming that the strategy of look-ahead controllers depends on the prediction quality of the vehicle model, the actions of these controllers should alter if $\mathcal{M}_4$-RRLM is used in conjunction with the controller. The performance of controllers might improve due to the improved prediction quality of $\mathcal{M}_4$-RRLM, compared with the conventional abc white-box model.

Generally, robust and regularized or wind-up stable estimators worked superior compared to standard estimators within Chapter 5. The used real world data is challenging due to outliers and periods of poor excitation. The multi-input-single-output output-error model is superior in prediction of the vehicle tractive force, whereas the random-walk errors-in-variables model convinces in vehicle mass estimation. The best model estimator combination for vehicle mass estimation was $\mathcal{M}_4$ with the Stenlund-Gustafsson IV M-Kalman filter (SGIVMKF). Conversely to the results herein, all cited related work used output-error models for parameter estimation. There is no model selection reported in vehicle parameter estimation literature elsewhere, which might give a reason why the random-walk errors-in-variables model has not been considered.

6.2 Contributions

Another heading for this section could be why was it worth to write this dissertation or why should you read it?

This dissertation applies major concepts from system identification such as model selection and model validation on a vehicle state prediction problem. The common way in vehicle science is to define some model, apply a favorite estimator, and measure the model quality in terms of the parameter error. Research with variation in the model structure and estimator is rare.

The main contribution is to find in Chapter 3 which attempts to provide a well organized survey of linear model structures and estimators. This survey begins with well applied standard estimators and shows connections to more sophisticated ones. Two concepts are emphasized.

The first is robustness. Robust estimators are from my point of view preferable, if they are easy to understand and not more challenging with respect to implementation and computational effort compared with non-robust estimators. The M-estimator gives an ideal concept to alter non-robust methods into robust ones. Actually, the modifications which lead from RLS into recursive M-estimator (RLM) are small. However, this small modification makes a huge difference if there are outliers in the data. The phrase: *if there are outliers in the data?* is misleading. Nowadays complex systems require more the question *are we confident enough to assume that there are no outliers?* The drawback of loosing optimality in RLM compared with

RLS (see end of Section 3.5.4) was from my professional experience always minor compared with the advantages the robust estimator offers.
The second concept which characterizes this dissertation is regularization. The record of real world data indicates that rich excitation was not ensured at all times. Hence, regularized or wind-up stable estimators played an important role within this work. Similar to robustness, the required modifications in the estimators, to include regularization or wind-up stability, are maintainable.
The next contribution of this work has to do with reproducibility which was discussed in Section 1.4. Common practice in engineering is to present nice results and to show *that it works fine*. This dissertation makes no difference in this matters. It is quite impossible to describe engineer's work to such an extend that full reproducibility is ensured and Chapter 5 is an example for this issue. However, this issue gave motivation to make at least some key contributions reusable. Accordingly, all discussed estimators are presented as algorithms and available as Matlab code in the additional material of this work (Chapter 6.3). The simulation experiments in Chapter 3 might motivate practitioners to apply some of the presented algorithms on own problems.
A contribution for vehicle look-ahead control applications is the result from model selection in Chapter 5, which gives arguments to use gray-box models instead of white-box models for vehicle tractive force prediction.

6.3 Open problems

The list of open problems which were marked with the ✎ symbol on page margins is quite long. First, we will focus on topics which are of general interest in system identification and may motivate further research. Accordingly, the following paragraphs result from the discussion in Chapter 3.
First, recursive GTLS (RGTLS) needs robustness and regularization to produce meaningful estimates when the data contains outliers and periods of

poor excitation. As regularization is a topic of interest in total least squares (TLS) research, a recursive regularized estimator might offer a fruitful research topic.

Second, a TLS-based solution of the random-walk errors-in-variables model would give a powerful errors-in-variables estimator, which comprises RGTLS as special case. Connections to robust Kalman filters are quite likely. Hence, a TLS-based solution of the random-walk errors-in-variables model would fit nicely into related research and might improve the performance of Kalman filters or generalize them.

Third, a throughout study of the presented polynomial Kalman smoother (PKS) with more applications would offer rich content for research. Extensions of PKS such as robust and regularized versions or an errors-in-variables PKS are desirable.

The second category of open topics are more vehicle specific and arise from the discussion in Chapter 5.

First, the test runs were made arbitrarily and data was recorded as it came. Hence, it cannot be stated that all relevant driving styles are covered within the data record. All ends up in the question *how much testing is enough testing*, which needs further investigation.

Second, more research is needed to check if more sophisticated models of the gradient angle and path radius give better accuracy or influence the accuracy of the vehicle tractive force gray-box model noticeably.

Third, a study of interacting multiple model estimation (IMME) with applications in vehicle tractive force prediction might offer additional accuracy in the prediction of the vehicle tractive force.

Fourth, in accordance with the prior open topic arises the question: *how to choose an optimal regularization parameter in a recursive fashion for vehicle tractive force estimation.* This point might also fit in the first list of general system identification topics if we shrink the question into: *optimal recursive regularization parameter.*

Fifth, an application where a vehicle look-ahead controller is used in parallel with a gray-box model might give enhanced energy efficient driving strategies.

Summary

This chapter summarized the main conclusions of this work and explained the main contributions. A lists of open topics in system identification, and open topics in related vehicle parameter estimation and control applications close this chapter and this dissertation. Let me add as final remark that I am open for collaborative research, so please let me know.

Algorithms

Outline

This chapter gives a list of algorithms from Chapter 3 and connects these algorithms with the implementation in Matlab® that is available as supplementary material of this dissertation. Please download the supplementary material from the public git repository:
https://bitbucket.org/Stephan_Rhode/dissertation_code.
I am grateful for comments, spotted bugs and suggested enhancements. Please feel free to create issues directly on the bitbucket repository page. Most algorithms of Chapter 3 are included in a single function which is called `fcn_LKFParaEst.m`. This function requires several inputs to compute parameter estimates. The following overview shows how to call this function, whereas we focus on the input structure that is called type and the instruments (called Aiv inside the function), because all other input arguments of the function are self-explanatory.

Algorithm 3.1 (RLS)

The function `fcn_LKFParaEst.m` resambles Algorithm 3.1 (RLS) by setting type.: FilterType-'RLS', CostFcn-'L2', robScale-'Off', R-λ, and Aiv-A.

Algorithm 3.2 (RLM)

Adjust the function inputs to type.: FilterType-'RLS', CostFcn-'Huber', robScale -'Off', R-λ, and Aiv-A to resamble Algorithm 3.2 (RLM). Note that RLM can also be used with other influence functions that were not shown within this work. The available options for CostFcn are: 'L1', 'L2' (non-robust), 'L1-L2', 'Huber', 'Cauchy', 'alphaDetector', 'Hampel', 'Tukey' and 'Myriad'.

Algorithm 3.3 (RLM with robust estimated scale)

See the previous setup for Algorithm 3.2 (RLM) and change type.robScale from 'Off' into 'On'.

Algorithm 3.4 (RRLM)

Adjust the function inputs to type.: FilterType-'LevenbergMarquardtRLS' or FilterType-'TikhonovRLS', CostFcn-'Huber', robScale-'Off', R-λ, and Aiv-A to resamble Algorithm 3.4 (RRLM). Note that RRLM can also be used with other influence functions.

Algorithm 3.9 (RIV)

The function `fcn_LKFParaEst.m` resambles Algorithm 3.9 (RIV) by setting type.: FilterType-'RLS', CostFcn-'L2', robScale-'Off', R-λ, and Aiv-$\mathfrak{A}$.

Algorithm 3.10 (RIVM)

Adjust the function inputs to type.: FilterType-'RLS', CostFcn-'Huber', robScale -'Off', R-λ, and Aiv-$\mathfrak{A}$ to resamble Algorithm 3.10 (RIVM). Also, numerous other influence functions can be used.

Algorithm 3.11 (RIVM with robust estimated scale)

See the previous setup for Algorithm 3.10 (RIVM) and switch type.robScale from 'Off' into 'On'.

Algorithm 3.12 (RRIVM)

Adjust the function inputs to type.: FilterType-'LevenbergMarquardtRLS' or FilterType-'TikhonovRLS', CostFcn-'Huber', robScale-'Off', R-λ, and Aiv-$\mathfrak{A}$ to resamble Algorithm 3.12 (RRIVM). Remember the available other influence functions.

Algorithm 3.13 (KF) – Algorithm 3.20 (SGIVMKF)

I assume that the reader is now familiar how to call the function to compute parameter estimates with a desired estimator (algorithm). The same concept holds for the Kalman filters, which were shown in Algorithm 3.13 (KF) – Algorithm 3.20 (SGIVMKF). However, the function call requires here type.Q-Q or P_d as additional input.

Summary

This chapter explained how to use a Matlab function from the supplementary material. The required input arguments for `fcn_LKFParaEst.m` are explained according to a variety of different estimators.

References

[1] DIN 70000. *Fahrzeugdynamik und Fahrverhalten*. Hrsg. von Normenausschuß Kraftfahrzeuge (FAKRA). DIN Deutsches Institut für Normung e.V., 1994.

[2] J. Agulló. „New algorithms for computing the least trimmed squares regression estimator". In: *Computational Statistics & Data Analysis* 36.4 (2001), S. 425–439. DOI: 10.1016/S0167-9473(00)00056-6.

[3] M.A. Akella und K. Subbarao. „A novel parameter projection mechanism for smooth and stable adaptive control". In: *Systems & Control Letters* 54.1 (2005), S. 43–51. DOI: 10.1016/j.sysconle.2004.06.004.

[4] B.M. Åkesson, J.B. Jørgensen, N.K. Poulsen und S.B. Jørgensen. „A generalized autocovariance least-squares method for Kalman filter tuning". In: *Journal of Process Control* 18.7-8 (2008), S. 769–779. DOI: 10.1016/j.jprocont.2007.11.003.

[5] F. Albu. „Improved Variable Forgetting Factor Recursive Least Square Algorithm". In: *Proc. of ICARCV.* 2012.

[6] S. Altmannshofer. Personal communication. Zentrum für Angewandte Forschung der Technischen Hochschule Ingolstadt. 17. Okt. 2014.

[7] R. Arablouei und K. Dogancay. „Linearly-Constrained Recursive Total Least-Squares Algorithm". In: *Signal Processing Letters, IEEE* 19.12 (Dez. 2012), S. 821–824. DOI: 10.1109/LSP.2012.2221705.

[8] A.Y. Aravkin, B.M. Bell, J.V. Burke und G. Pillonetto. „An l_1 -Laplace Robust Kalman Smoother". In: *Automatic Control, IEEE Transactions on* 56.12 (2011), S. 2898–2911. DOI: 10.1109/TAC.2011.2141430.

[9] A Assa und F. Janabi-Sharifi. „A Robust Vision-Based Sensor Fusion Approach for Real-Time Pose Estimation". In: *Cybernetics, IEEE Transactions on* 44.2 (Feb. 2014), S. 217–227. DOI: 10.1109/TCYB.2013.2252339.

[10] Karl J Åström und Björn Wittenmark. *Adaptive control.* Courier Corporation, 2013.

[11] Audi. *http://www.audi-technology-portal.de/en/multimedia_en/predictive-route-data.* webpage. accessed on 2014/11/17. 2104.

[12] M. Back. „Prädiktive Antriebsregelung zum energieoptimalen Betrieb von Hybridfahrzeugen“. Diss. Universität Karlsruhe (TH), 2005. DOI: 10.5445/KSP/1000004546.

[13] H.S. Bae, J. Ryu und J.C. Gerdes. „Road grade and vehicle parameter estimation for longitudinal control using GPS“. In: *IEEE Conference on Intelligent Transportation Systems, Proceedings, ITSC*. 2001, S. 166–171. DOI: 10.1109/ITSC.2001.948649.

[14] E. Bai. „A random least-trimmed-squares identification algorithm“. In: *Automatica* 39.9 (2003), S. 1651–1659. DOI: 10.1016/S0005-1098(03)00193-6.

[15] Y. Bar-Shalom, X.-R. Li und T. Kirubarajan. *Estimation with applications to tracking and navigation*. John Wiley & Sons, 2001. DOI: 10.1002/0471221279.

[16] V.A. Bavdekar, A.P. Deshpande und S.C. Patwardhan. „Identification of process and measurement noise covariance for state and parameter estimation using extended Kalman filter“. In: *Journal of Process Control* 21.4 (2011), S. 585–601. DOI: 10.1016/j.jprocont.2011.01.001.

[17] R.E. Bellman und S.E. Dreyfus. *Applied dynamic programming*. Princeton University Press, 1962.

[18] H.A.P. Blom und Y. Bar-Shalom. „The interacting multiple model algorithm for systems with Markovian switching coefficients“. In: *Automatic Control, IEEE Transactions on* 33.8 (Aug. 1988), S. 780–783. DOI: 10.1109/9.1299.

[19] Bosch. *http://rb-aa.bosch.com/automotive/catalogues/de/de/start/media/sensoren_fuer_sonderkunden_industriekd.pdf*. webpage. accessed on 2014/10/11. 2014.

[20] G.E.P. Box und N.R Draper. *Empirical model-building and response surfaces*. John Wiley & Sons, 1987.

[21] K. Brabanter, K. Pelckmans, J. Brabanter, M. Debruyne, J.A.K. Suykens, M. Hubert und B. Moor. „Robustness of Kernel Based Regression: A Comparison of Iterative Weighting Schemes“. In: *Artificial Neural Networks – ICANN 2009*. Hrsg. von C. Alippi, M. Polycarpou, C. Panayiotou und G. Ellinas. Bd. 5768. Lecture Notes in Computer Science. Springer Berlin Heidelberg, 2009, S. 100–110. DOI: 10.1007/978-3-642-04274-4_11.

[22] M. Brand. „Incremental Singular Value Decomposition of Uncertain Data with Missing Values“. In: *Computer Vision-ECCV 2002*. Springer, 2002, S. 707–720. DOI: 10.1007/3-540-47969-4_47.

[23] M. Brand. „Fast low-rank modifications of the thin singular value decomposition“. In: *Linear algebra and its applications* 415.1 (2006), S. 20–30. DOI: 10.1016/j.laa.2005.07.021.

[24] F.T. Buckley. „ABCD-An Improved Coast Down Test and Analysis Method“. In: *SAE transactions* 104 (1995), S. 1136–1148. DOI: 10.4271/950626.

[25] Y. Bulut, D. Vines-Cavanaugh und D. Bernal. „Process and Measurement Noise Estimation for Kalman Filtering“. In: *Structural Dynamics, Volume 3.* Hrsg. von T. Proulx. Conference Proceedings of the Society for Experimental Mechanics Series. Springer New York, 2011, S. 375–386. DOI: 10.1007/978-1-4419-9834-7_36.

[26] M. Burckhardt. *Fahrwerktechnik: Radschlupf-Regelsysteme.* 1. Aufl. Vogel-Fachbuch : Kraftfahrzeugtechnik Fahrwerktechnik. Würzburg: Vogel, 1993.

[27] K.P. Burnham und D.R. Anderson. *Model selection and multimodel inference: a practical information theoretic approach.* 2. ed. New York; Berlin; Heidelberg [u.a.]: Springer, 2002, XXVI, 488 S.

[28] K.P. Burnham und D.R. Anderson. „Multimodel inference understanding AIC and BIC in model selection“. In: *Sociological methods & research* 33.2 (2004), S. 261–304. DOI: 10.1177/0049124104268644.

[29] T. Bylander und B. Rosen. „A perceptron-like online algorithm for tracking the median“. In: *Neural Networks,1997., International Conference on.* Bd. 4. 1997, S. 2219–2224. DOI: 10.1109/ICNN.1997.614292.

[30] L. Cao und H. Schwartz. „A directional forgetting algorithm based on the decomposition of the information matrix“. In: *Automatica* 36.11 (2000), S. 1725–1731. DOI: 10.1016/S0005-1098(00)00093-5.

[31] L. Cao und H.M. Schwartz. „A novel recursive algorithm for directional forgetting“. In: *American Control Conference, 1999. Proceedings of the 1999.* Bd. 2. Juni 1999, S. 1334–1338. DOI: 10.1109/ACC.1999.783584.

[32] B. Carew und P. Belanger. „Identification of optimum filter steady-state gain for systems with unknown noise covariances“. In: *Automatic Control, IEEE Transactions on* 18.6 (Dez. 1973), S. 582–587. DOI: 10.1109/TAC.1973.1100420.

[33] R. de Castro, R.E. Araujo und D. Freitas. „Real-time estimation of tyre-road friction peak with optimal linear parameterisation“. In: *Control Theory Applications, IET* 6.14 (2012), S. 2257–2268. DOI: 10.1049/iet-cta.2011.0424.

[34] S.C. Chan, Y.J. Chu und Z.G. Zhang. „A new regularized QRD recursive least M-estimate algorithm: Performance analysis and applications“. In: *Green Circuits and Systems (ICGCS), 2010 International Conference on.* Juni 2010, S. 190 –195. DOI: 10.1109/ICGCS.2010.5543069.

[35] T.F. Chan, G.H. Golub und R.J. LeVeque. „Algorithms for Computing the Sample Variance: Analysis and Recommendations“. In: *The American Statistician* 37.3 (1983), S. 242–247. URL: http://www.jstor.org/stable/2683386.

[36] V. Chandola, A. Banerjee und V. Kumar. „Anomaly Detection: A Survey". In: *ACM Comput. Surv.* 41.3 (Juli 2009), 15:1–15:58. DOI: 10.1145/1541880.1541882.

[37] G. Cirrincione und M. Cirrincione. *Neural Based Orthogonal Data Fitting: The EXIN Neural Networks.* Bd. 38. Wiley, 2010. DOI: 10.1002/9780470638286.

[38] V.J Clancey. „Statistical methods in chemical analyses". In: *Nature* 159 (1947), S. 339–340. DOI: 10.1038/159339c0.

[39] H. Dahmani, M. Chadli, A. Rabhi und A. El Hajjaji. „Road curvature estimation for vehicle lane departure detection using a robust Takagi-Sugeno fuzzy observer". In: *Vehicle System Dynamics* 51.5 (2013), S. 581–599. DOI: 10.1080/00423114.2011.642806.

[40] H. Dai und N.K. Sinha. „Robust recursive least-squares method with modified weights for bilinear system identification". In: *Control Theory and Applications, IEE Proceedings D* 136.3 (Mai 1989), S. 122–126.

[41] C.E. Davila. „An efficient recursive total least squares algorithm for FIR adaptive filtering". In: *Signal Processing, IEEE Transactions on* 42.2 (Feb. 1994), S. 268–280. DOI: 10.1109/78.275601.

[42] S. De Bruyne, H. Van der Auweraer, P. Diglio und J. Anthonis. „Online Estimation of Vehicle Inertial Parameters for Improving Chassis Control Systems". In: *Proceedings of the 18th World Congress, The International Federation of Automatic Control.* Bd. 18. 1. 2011, S. 1814–1819. DOI: 10.3182/20110828-6-IT-1002.03379.

[43] N. Denis, M.R. Dubois, K.A. Gil, T. Driant und A. Desrochers. „Range prediction for a three-wheel plug-in hybrid electric vehicle". In: *Transportation Electrification Conference and Expo (ITEC), 2012 IEEE.* Juni 2012, S. 1–6. DOI: 10.1109/ITEC.2012.6243415.

[44] F. Ding, T. Chen und L. Qiu. „Bias compensation based recursive least-squares identification algorithm for MISO systems". In: *Circuits and Systems II: Express Briefs, IEEE Transactions on* 53.5 (Mai 2006), S. 349–353. DOI: 10.1109/TCSII.2005.862281.

[45] P.S.R. Diniz. *Adaptive filtering: algorithms and practical implementation.* 4. Aufl. Springer, 2008. DOI: 10.1007/978-0-387-68606-6.

[46] Dspace. *http://www.dspace.com/en/inc/home/products/hw/micautob.cfm.* webpage. accessed on 2014/09/19. 2014.

[47] Z.M. Durovic und B.D. Kovacevic. „Robust estimation with unknown noise statistics". In: *Automatic Control, IEEE Transactions on* 44.6 (1999), S. 1292–1296. DOI: 10.1109/9.769393.

[48] A. Einstein. „Physik und Realität". In: *Journal of the Franklin Institute* 221.3 (1936), S. 313–347. DOI: 10.1016/S0016-0032(36)91045-1.

[49] L. El Ghaoui und H. Lebret. „Robust Solutions to Least-Squares Problems with Uncertain Data". In: *SIAM Journal on Matrix Analysis and Applications* 18.4 (1997), S. 1035–1064. DOI: 10.1137/S0895479896298130.

[50] M. Evestedt und A. Medvedev. „Stationary behavior of an anti-windup scheme for recursive parameter estimation under lack of excitation". In: *Automatica* 42.1 (2006), S. 151–157. DOI: 10.1016/j.automatica.2005.08.015.

[51] M. Evestedt und A. Medvedev. „Recursive parameter estimation by means of the SG-algorithm". In: *17th IFAC world congress.* 2008.

[52] H.K. Fathy, D. Kang und J.L. Stein. „Online vehicle mass estimation using recursive least squares and supervisory data extraction". In: *American Control Conference, 2008.* 2008, S. 1842–1848. DOI: 10.1109/ACC.2008.4586760.

[53] D.-Z. Feng und W.X. Zheng. „Fast approximate inverse power iteration algorithm for adaptive total least-squares FIR filtering". In: *IEEE Transactions on Signal Processing* 54.10 (Okt. 2006), S. 4032–4039. DOI: 10.1109/TSP.2006.880245.

[54] D.-Z. Feng und W.X. Zheng. „Recursive total instrumental-variable algorithm for solving over-determined normal equations and its applications". In: *Signal Processing* 87.5 (2007), S. 918–936. DOI: 10.1016/j.sigpro.2006.09.015.

[55] D.-Z. Feng, X.-D. Zhang, D.-X. Chang und W.X. Zheng. „A fast recursive total least squares algorithm for adaptive FIR filtering". In: *Signal Processing, IEEE Transactions on* 52.10 (Okt. 2004), S. 2729–2737. DOI: 10.1109/TSP.2004.834260.

[56] T.R. Fortescue, L.S. Kershenbaum und B.E. Ydstie. „Implementation of self-tuning regulators with variable forgetting factors". In: *Automatica* 17.6 (1981), S. 831–835. DOI: 10.1016/0005-1098(81)90070-4.

[57] B. Friedlander. „The overdetermined recursive instrumental variable method". In: *Automatic Control, IEEE Transactions on* 29.4 (Apr. 1984), S. 353–356. DOI: 10.1109/TAC.1984.1103531.

[58] M. Frigge, D.C. Hoaglin und B. Iglewicz. „Some Implementations of the Boxplot". In: *The American Statistician* 43.1 (1989), S. 50–54. DOI: 10.2307/2685173.

[59] B. Ganji, A.Z. Kouzani und H. Khayyam. „Look-ahead intelligent energy management of a parallel hybrid electric vehicle". In: *Fuzzy Systems (FUZZ), 2011 IEEE International Conference on.* Juni 2011, S. 2335–2341. DOI: 10.1109/FUZZY.2011.6007495.

[60] E. Garfield. „What is the primordial reference for the phrase ‚publish or perish'?" In: *The Scientist* 10.12 (1996), S. 11.

[61] O. Gietelink, J. Ploeg, B. De Schutter und M. Verhaegen. „Development of advanced driver assistance systems with vehicle hardware-in-the-loop simulations“. In: *Vehicle System Dynamics* 44.7 (2006), S. 569–590. DOI: 10.1080/00423110600563338.

[62] G.H. Golub und C.F. Van Loan. „An analysis of the total least squares problem“. In: *SIAM Journal on Numerical Analysis* 17 (1980), S. 883–893.

[63] G.H. Golub und C.F. Van Loan. *Matrix computations*. Bd. 3. Johns Hopkins Univ Pr, 1996.

[64] J.G. Gonzalez und G.R. Arce. „Statistically-efficient Filtering in Impulsive Environments: Weighted Myriad Filters“. In: *EURASIP J. Appl. Signal Process.* 2002.1 (Jan. 2002), S. 4–20. DOI: 10.1155/S1110865702000483.

[65] Kwai S. Goodwin Graham C. ; Sin. *Adaptive filtering prediction and control*. Prentice-Hall information and system sciences series. Englewood Cliffs, NJ: Prentice-Hall, 1984.

[66] K.S. Grewal und P.M. Darnell. „Model-based EV range prediction for Electric Hybrid Vehicles“. In: *Hybrid and Electric Vehicles Conference 2013 (HEVC 2013), IET*. Nov. 2013, S. 1–6. DOI: 10.1049/cp.2013.1895.

[67] M. Gu und S.C. Eisenstat. „A stable and fast algorithm for updating the singular value decomposition“. In: *New Haven: Yale University Department of Computer Science. RR-939* (1993).

[68] S. Gunnarsson. „Combining tracking and regularization in recursive least squares identification“. In: *Decision and Control, 1996., Proceedings of the 35th IEEE Conference on*. Bd. 3. Dez. 1996, 2551–2552 vol.3. DOI: 10.1109/CDC.1996.573481.

[69] K.L. Haken. *Grundlagen der Kraftfahrzeugtechnik*. 2. Aufl. Fahrzeugtechnik. München: Hanser, 2011.

[70] C. Halfmann und H. Holzmann. *Adaptive Modelle für die Kraftfahrzeugdynamik*. Berlin: Springer, 2003.

[71] K.J. Han, I.K. Kim, H.Y. Jo und K.S. Huh. „Development and experimental evaluation of an online estimation system for vehicle mass“. In: *Proceedings of the Institution of Mechanical Engineers, Part D: Journal of Automobile Engineering* 223.2 (2009), S. 167–177. DOI: 10.1243/09544070JAUTO991.

[72] O. Hartmann, R. Schweiger, R. Wagner, F. Schule, M. Gabb und K. Dietmayer. „Night time road curvature estimation based on Convolutional Neural Networks“. In: *Intelligent Vehicles Symposium (IV), 2013 IEEE*. Juni 2013, S. 841–846. DOI: 10.1109/IVS.2013.6629571.

[73] S. Haykin. *Neural networks and learning machines.* 3. ed., internat. ed. Upper Saddle River: Pearson, 2009.

[74] B. Heißing. *Fahrwerkhandbuch : Grundlagen, Fahrdynamik, Komponenten, Systeme, Mechatronik, Perspektiven.* Hrsg. von M. Esroy und S. Gies. 3. SpringerLink : Bücher. Wiesbaden: Vieweg+Teubner Verlag / Springer Fachmedien Wiesbaden GmbH, Wiesbaden, 2011. DOI: 10.1007/978-3-8348-8168-7.

[75] E. Hellström, M. Ivarsson, J. Åslund und L. Nielsen. „Look-ahead control for heavy trucks to minimize trip time and fuel consumption". In: *Control Engineering Practice* 17.2 (2009), S. 245–254. DOI: 10.1016/j.conengprac.2008.07.005.

[76] V.J. Hodge und J. Austin. „A Survey of Outlier Detection Methodologies". In: *Artificial Intelligence Review* 22.2 (2004), S. 85–126. DOI: 10.1007/s10462-004-4304-y.

[77] M. Hofmann, C. Gatu und E.J. Kontoghiorghes. „An Exact Least Trimmed Squares Algorithm for a Range of Coverage Values". In: *Journal of Computational and Graphical Statistics* 19.1 (2010), S. 191–204. DOI: 10.1198/jcgs.2009.07091.

[78] S. Hong, T. Smith, F. Borrelli und J.K. Hedrick. „Vehicle inertial parameter identification using Extended and unscented Kalman Filters". In: *Intelligent Transportation Systems - (ITSC), 2013 16th International IEEE Conference on.* Okt. 2013, S. 1436–1441. DOI: 10.1109/ITSC.2013.6728432.

[79] S. Hong, C. Lee, F. Borrelli und J.K. Hedrick. „A Novel Approach for Vehicle Inertial Parameter Identification Using a Dual Kalman Filter". In: *Intelligent Transportation Systems, IEEE Transactions on* PP.99 (2014), S. 1–11. DOI: 10.1109/TITS.2014.2329305.

[80] J. Howell, C. Sherwin, MA Passmore und GM Le Good. „Aerodynamic drag of a compact SUV as measured on-road and in the wind tunnel". In: *SAE Transactions, Journal of Passenger Cars - Mechanical Systems* 1 (2002), S. 583–590. DOI: 10.4271/2002-01-0529.

[81] P.J. Huber. „Robust Estimation of a Location Parameter". In: *The Annals of Mathematical Statistics* 35.1 (1964), S. 73–101. DOI: 10.1214/aoms/1177703732.

[82] P.J. Huber und E.M. Ronchetti. *Robust statistics.* 2. Aufl. Wiley series in probability and statistics. Hoboken, N.J: Wiley, 2009. DOI: 10.1002/9780470434697.

[83] K. Huh, S. Lim, J. Jung, D. Hong, S. Han, K. Han, H. Young und JM Yun. „Vehicle mass estimator for adaptive roll stability control". In: *SAE Technical papers* (2007), S. 04–16. DOI: 10.4271/2007-01-0820.

[84] K. Hyniova, A. Stribrsky, J. Honcu und A. Kruczek. „Active suspension system with linear electric motor". In: *WSEAS Transactions on Systems* 8.2 (2009), S. 278–287.

[85] R. Isermann, Hrsg. *Fahrdynamik-Regelung : Modellbildung, Fahrerassistenzsysteme, Mechatronik.* Wiesbaden: Friedr. Vieweg & Sohn Verlag / GWV Fachverlage GmbH, Wiesbaden, 2006. DOI: 10.1007/978-3-8348-9049-8.

[86] R. Isermann und M. Münchhof. *Identification of dynamic systems : an introduction with applications.* Heidelberg: Springer, 2011. DOI: 10.1007/978-3-540-78879-9_1.

[87] J. Jiang und R. Cook. „Fast parameter tracking RLS algorithm with high noise immunity". In: *Electronics Letters* 28.22 (Okt. 1992), S. 2043–2045. DOI: 10.1049/el:19921309.

[88] K. Jo, K. Chu und M. Sunwoo. „Interacting Multiple Model Filter-Based Sensor Fusion of GPS With In-Vehicle Sensors for Real-Time Vehicle Positioning". In: *Intelligent Transportation Systems, IEEE Transactions on* 13.1 (März 2012), S. 329–343. DOI: 10.1109/TITS.2011.2171033.

[89] M. Karasalo und X. Hu. „An optimization approach to adaptive Kalman filtering". In: *Automatica* 47.8 (2011), S. 1785–1793. DOI: 10.1016/j.automatica.2011.04.004.

[90] H. Khayyam, S. Nahavandi, E. Hu, A. Kouzani, A. Chonka, J. Abawajy, V. Marano und S. Davis. „Intelligent energy management control of vehicle air conditioning via look-ahead system". In: *Applied Thermal Engineering* 31.16 (2011), S. 3147–3160. DOI: 10.1016/j.applthermaleng.2011.05.023.

[91] U. Kiencke und L. Nielsen. *Automotive control systems : for engine, driveline, and vehicle.* 2. Aufl. Berlin: Springer, 2005. DOI: 10.1007/b137654.

[92] Kistler. *http://www.corrsys-datron.com/optical_sensors.htm.* webpage. accessed on 2014/10/12. 2014.

[93] E.M. Knorr und R.T. Ng. „Algorithms for mining distance-based outliers in large datasets". In: *Proceedings of the International Conference on Very Large Data Bases.* 1998, S. 392–403.

[94] D. Kubus, T. Kroger und F.M. Wahl. „On-line estimation of inertial parameters using a recursive total least-squares approach". In: *Intelligent Robots and Systems, 2008. IROS 2008. IEEE/RSJ International Conference on.* Sep. 2008, S. 3845–3852. DOI: 10.1109/IROS.2008.4650672.

[95] J. Lampe. „Solving regularized total least squares problems based on eigenproblems". Diss. Technische Universität Hamburg-Harburg, 2010. URL: http://doku.b.tu-harburg.de/volltexte/2010/959.

[96] P. Lemmerling und B. De Moor. „Misfit versus latency". In: *Automatica* 37.12 (2001), S. 2057–2067. DOI: 10.1016/S0005-1098(01)00180-7.

[97] S.H. Leung und C.F. So. „Gradient-based variable forgetting factor RLS algorithm in time-varying environments“. In: *Signal Processing, IEEE Transactions on* 53.8 (Aug. 2005), S. 3141–3150. DOI: 10.1109/TSP.2005.851110.

[98] Yi. Liang und O.M. Kvalheim. „Robust methods for multivariate analysis – a tutorial review“. In: *Chemometrics and Intelligent Laboratory Systems* 32.1 (1996), S. 1–10. DOI: 10.1016/0169-7439(95)00006-2.

[99] J. Lim, N. Choi und K. Sung. „Robust Recursive TLS (Total Least Square) Method Using Regularized UDU Decomposed for FNN (Feedforward Neural Network) Training“. In: *Advances in Neural Networks*. Springer, 2005, S. 133–137. DOI: 10.1007/11427391_92.

[100] L. Ljung. *System identification : theory for the user.* Prentice-Hall information and system sciences series. Upper Saddle River, NJ: Prentice Hall, 1999.

[101] L. Ljung. *System Identification Toolbox™: User's Guide.* Mathworks. 2012.

[102] L. Ljung und S. Gunnarsson. „Adaptation and tracking in system identification : A survey“. In: *Automatica* 26.1 (1990), S. 7–21. DOI: 10.1016/0005-1098(90)90154-A.

[103] S. Lu, S.V. Pereverzev und U. Tautenhahn. „A model function method in regularized total least squares“. In: *Applicable Analysis* 89.11 (2010), S. 1693–1703. DOI: 10.1080/00036811.2010.492502.

[104] H.H. Madden. „Comments on the Savitzky-Golay convolution method for least-squares-fit smoothing and differentiation of digital data“. In: *Analytical Chemistry* 50.9 (1978), S. 1383–1386. DOI: 10.1021/ac50031a048.

[105] I. Markovsky, D.M. Sima und S. Van Huffel. „Total least squares methods“. In: *Wiley Interdisciplinary Reviews: Computational Statistics* 2.2 (2010), S. 212–217. DOI: 10.1002/wics.65.

[106] I. Markovsky und S. Van Huffel. „Overview of total least-squares methods“. In: *Signal Processing* 87.10 (2007), S. 2283–2302. DOI: 10.1016/j.sigpro.2007.04.004.

[107] I. Markovsky, M.L. Rastello, A. Premoli, A. Kukush und S. Van Huffel. „The element-wise weighted total least-squares problem“. In: *Computational Statistics & Data Analysis* 50.1 (2006), S. 181–209. DOI: 10.1016/j.csda.2004.07.014.

[108] I. Markovsky, K. Usevich, M. Ishteva und S. Rhode. personal communication. department ELEC, Vrije Universiteit Brussel (VUB), Okt. 2013.

[109] W. Mayer. „Bestimmung und Aufteilung des Fahrwiderstandes im realen Fahrbetrieb“. Diss. Renningen: Universität Stuttgart, 2006.

[110] W. Mayer und J. Wiedemann. „Road load determination based on driving-torque-measurement“. In: *SAE transactions* 112.6 (2003), S. 941–950. DOI: 10.4271/2003-01-0933.

[111] R. McGill, J.W. Tukey und W.A. Larsen. „Variations of Box Plots“. In: *The American Statistician* 32.1 (1978), S. 12–16. DOI: 10.2307/2683468.

[112] M.L. McIntyre, T.J. Ghotikar, A. Vahidi, X. Song und D.M. Dawson. „A Two-Stage Lyapunov-Based Estimator for Estimation of Vehicle Mass and Road Grade“. In: *Vehicular Technology, IEEE Transactions on* 58.7 (Sep. 2009), S. 3177–3185. DOI: 10.1109/TVT.2009.2014385.

[113] R.K. Mehra. „On the identification of variances and adaptive Kalman filtering“. In: *Automatic Control, IEEE Transactions on* 15.2 (1970), S. 175–184. DOI: 10.1109/TAC.1970.1099422.

[114] M. Mitschke und H. Wallentowitz. *Dynamik der Kraftfahrzeuge.* 4. Aufl. VDI. Berlin: Springer, 2004.

[115] A. H. Mohamed und K. P. Schwarz. „Adaptive Kalman Filtering for INS/GPS“. In: *Journal of Geodesy* 73.4 (1999), S. 193–203. DOI: 10.1007/s001900050236.

[116] M. Moonen. „Systolic algorithms for recursive total least squares parameter estimation and mixed RLS/RTLS problems“. In: *International Journal of High Speed Electronics* 4 (1993), S. 55–55.

[117] K. Myers und B.D. Tapley. „Adaptive sequential estimation with unknown noise statistics“. In: *Automatic Control, IEEE Transactions on* 21.4 (Aug. 1976), S. 520–523. DOI: 10.1109/TAC.1976.1101260.

[118] K. Nam, S. Oh, H. Fujimoto und Y. Hori. „Estimation of Sideslip and Roll Angles of Electric Vehicles Using Lateral Tire Force Sensors Through RLS and Kalman Filter Approaches“. In: *Industrial Electronics, IEEE Transactions on* 60.3 (2013), S. 988–1000. DOI: 10.1109/TIE.2012.2188874.

[119] M. Nentwig, R. Schieber und M. Miegler. „Hardware-in-the-Loop Testing of Advanced Driver Assistance Systems“. In: *ATZelektronik worldwide eMagazine* 6.4 (2011), S. 10–15. DOI: 10.1365/s38314-011-0034-5.

[120] G. Noriega und S. Pasupathy. „Adaptive estimation of noise covariance matrices in real-time preprocessing of geophysical data“. In: *Geoscience and Remote Sensing, IEEE Transactions on* 35.5 (Sep. 1997), S. 1146–1159. DOI: 10.1109/36.628782.

[121] B.J. Odelson, M.R. Rajamani und J.B. Rawlings. „A new autocovariance least-squares method for estimating noise covariances“. In: *Automatica* 42.2 (2006), S. 303–308. DOI: 10.1016/j.automatica.2005.09.006.

[122] H.B. Pacejka. *Tire and vehicle dynamics.* Hrsg. von I. Besselink. 3. Aufl. Oxford, UK: Butterworth-Heinemann, 2012. DOI: 10.1016/B978-0-08-097016-5.00003-6.

[123] C. Paleologu, J. Benesty und S. Ciochina. „A Robust Variable Forgetting Factor Recursive Least-Squares Algorithm for System Identification“. In: *Signal Processing Letters, IEEE* 15 (2008), S. 597–600. DOI: 10.1109/LSP.2008.2001559.

[124] J.E. Parkum, N.K. Poulsen und J. Holst. „Recursive forgetting algorithms“. In: *International Journal of Control* 55.1 (1992), S. 109–128. DOI: 10.1080/00207179208934228.

[125] M.A. Passmore und G.M. Le Good. „A detailed drag study using the coastdown method“. In: *SAE Paper 940420* (1994), S. 119–126. DOI: 10.4271/940420.

[126] I. Petros und B. Fidan. *Adaptive control tutorial.* Advances in design and control ; 11. Philadelphia, Pa.: Society for Industrial und Applied Mathematics, 2006.

[127] A. Picard, R.S. Davis, M. Gläser und K. Fujii. „Revised formula for the density of moist air (CIPM-2007)“. In: *Metrologia* 45.2 (2008), S. 149–155. DOI: 10.1088/0026-1394/45/2/004.

[128] Porsche. *http://www.porsche.com/international/models/panamera/panamera-turbo-s/.* webpage. accessed on 2014/09/19. 2014.

[129] Porsche. *http://www.porsche.com/international/models/panamera/panamera/.* webpage. accessed on 2014/09/19. 2014.

[130] S.J. Qin und T.A. Badgwell. „A survey of industrial model predictive control technology“. In: *Control Engineering Practice* 11.7 (2003), S. 733–764. DOI: 10.1016/S0967-0661(02)00186-7.

[131] T. Radke. „Energieoptimale Längsführung von Kraftfahrzeugen durch Einsatz vorausschauender Fahrstrategien“. Diss. Karlsruher Institut für Technologie, 2013. DOI: 10.5445/KSP/1000035819.

[132] M.R. Rajamani und J.B. Rawlings. „Estimation of the disturbance structure from data using semidefinite programming and optimal weighting“. In: *Automatica* 45.1 (2009), S. 142–148. DOI: 10.1016/j.automatica.2008.05.032.

[133] S. Rhode, F. Bleimund und F. Gauterin. „Recursive Generalized Total Least Squares with Noise Covariance Estimation“. In: *19th IFAC World Congress.* Bd. 19. International Federation of Automatic Control, 2014, S. 4637–4643. DOI: 10.3182/20140824-6-ZA-1003.01568.

[134] S. Rhode und F. Gauterin. „Vehicle mass estimation using a total least-squares approach“. In: *Intelligent Transportation Systems (ITSC), 2012 15th International IEEE Conference on.* Sep. 2012, S. 1584–1589. DOI: 10.1109/ITSC.2012.6338638.

[135] S. Rhode und F. Gauterin. „Online estimation of vehicle driving resistance parameters with recursive least squares and recursive total least squares“. In: *Intelligent Vehicles Symposium (IV), 2013 IEEE.* Juni 2013, S. 269–276. DOI: 10.1109/IVS.2013.6629481.

[136] S. Rhode, K. Usevich, I. Markovsky und F. Gauterin. „A Recursive Restricted Total Least-squares Algorithm“. In: *Signal Processing, IEEE Transactions on* 62.21 (Nov. 2014), S. 5652–5662. DOI: 10.1109/TSP.2014.2350959.

[137] P. Riekert und T.E. Schunck. „Zur Fahrmechanik des gummibereiften Kraftfahrzeugs“. In: *Ingenieur-Archiv* 11.3 (1940), S. 210–224. DOI: 10.1007/BF02086921.

[138] F. Rosenblatt. „The perceptron: a probabilistic model for information storage and organization in the brain“. In: *Psychological Review* 65.6 (1958), S. 386–408. DOI: 10.1037/h0042519.

[139] P.J. Rousseeuw. „Least Median of Squares Regression“. In: *Journal of the American Statistical Association* 79.388 (1984), S. 871–880. DOI: 10.1080/01621459.1984.10477105.

[140] P.J. Rousseeuw und G.W. Bassett. „The Remedian: A Robust Averaging Method for Large Data Sets“. In: *Journal of the American Statistical Association* 85.409 (1990), S. 97–104. DOI: 10.1080/01621459.1990.10475311.

[141] P.J. Rousseeuw und C. Croux. „Alternatives to the Median Absolute Deviation“. In: *Journal of the American Statistical Association* 88.424 (1993), S. 1273–1283. DOI: 10.1080/01621459.1993.10476408.

[142] P.J. Rousseeuw und A.M. Leroy. *Robust regression and outlier detection.* Wiley series in probability and mathematical statistics : Applied probability and statistics. Wiley, 1987. DOI: 10.1002/0471725382.

[143] P.J. Rousseeuw und K. Van Driessen. „A Fast Algorithm for the Minimum Covariance Determinant Estimator“. In: *Technometrics* 41.3 (1999), S. 212–223. DOI: 10.1080/00401706.1999.10485670.

[144] P.J Rousseeuw und K. Van Driessen. „Computing LTS regression for large data sets“. In: *Data mining and knowledge discovery* 12.1 (2006), S. 29–45. DOI: 10.1007/s10618-005-0024-4.

[145] S. Saelid, O. Egeland und B. Foss. „A solution to the blow-up problem in adaptive controllers“. In: *Modeling, Identification and Control* 6.1 (1985), S. 21–37. DOI: 10.4173/mic.1985.1.3.

[146] S. Saelid und B. Foss. „Adaptive controllers with a vector variable forgetting factor“. In: *Decision and Control, 1983. The 22nd IEEE Conference on.* Bd. 22. Dez. 1983, S. 1488–1494. DOI: 10.1109/CDC.1983.269785.

[147] P. Sahlholm und K. Henrik Johansson. „Road grade estimation for look-ahead vehicle control using multiple measurement runs“. In: *Control Engineering Practice* 18 (2009), S. 1328–1341. DOI: 10.1016/j.conengprac.2009.09.007.

[148] A. Savitzky und M.J.E. Golay. „Smoothing and Differentiation of Data by Simplified Least Squares Procedures.“ In: *Analytical Chemistry* 36.8 (1964), S. 1627–1639. DOI: 10.1021/ac60214a047.

[149] A.H. Sayed. „A framework for state-space estimation with uncertain models“. In: *Automatic Control, IEEE Transactions on* 46.7 (2001), S. 998–1013. DOI: 10.1109/9.935054.

[150] A.H. Sayed. *Fundamentals of adaptive filtering*. Hohoken, New Jersey: Wiley-IEEE Press, 2003.

[151] A.H. Sayed und T. Kailath. „A state-space approach to adaptive RLS filtering“. In: *Signal Processing Magazine, IEEE* 11.3 (1994), S. 18–60. DOI: 10.1109/79.295229.

[152] B. Schaffrin. „A note on constrained total least-squares estimation“. In: *Linear algebra and its applications* 417.1 (2006), S. 245–258. DOI: 10.1016/j.laa.2006.03.044.

[153] B. Schaffrin und A. Wieser. „On weighted total least-squares adjustment for linear regression“. In: *Journal of Geodesy* 82 (7 2008), S. 415–421. DOI: 10.1007/s00190-007-0190-9.

[154] J. Schoukens. *System Identification*. lecture notes. Apr. 2013.

[155] D. Schramm. *Modellbildung und Simulation der Dynamik von Kraftfahrzeugen*. Hrsg. von M. Hiller und R. Bardini. SpringerLink : Bücher. Berlin, Heidelberg: Springer-Verlag Berlin Heidelberg, 2010. DOI: 10.1007/978-3-540-89315-8.

[156] M. Schuermans, I. Markovsky, P. Wentzell und S. Van Huffel. „On the equivalence between total least squares and maximum likelihood PCA“. In: *Analytica Chimica Acta* 544 (2005), S. 254–267. DOI: 10.1016/j.aca.2004.12.059.

[157] S. Semmler. „Regelung der Fahrzeugbremsdynamik mit kontinuierlich einstellbaren Radbremsen“. Diss. TU Darmstadt, 2006.

[158] B. Seyfe und S. Valaee. „A new choice of penalty function for robust multiuser detection based on M-estimation“. In: *Communications, IEEE Transactions on* 53.2 (Feb. 2005), S. 224–227. DOI: 10.1109/TCOMM.2004.842001.

[159] Y. Shen, B. Li und Y. Chen. „An iterative solution of weighted total least-squares adjustment“. In: *Journal of Geodesy* 85 (4 2011), S. 229–238. DOI: 10.1007/s00190-010-0431-1.

[160] D. Sima. „Regularization Techniques in Model Fitting and Parameter Estimation“. Diss. K.U.Leuven, 2006.

[161] D. Simon. „Kalman filtering with state constraints: a survey of linear and nonlinear algorithms“. In: *Control Theory Applications, IET* 4.8 (Aug. 2010), S. 1303–1318. DOI: 10.1049/iet-cta.2009.0032.

[162] Dan Simon. *Optimal state estimation : Kalman, H_∞ and nonlinear approaches.* Hoboken, NJ: Wiley-Interscience, 2006.

[163] C.F. So, S.C. Ng und S.H. Leung. „Gradient based variable forgetting factor RLS algorithm“. In: *Signal Processing* 83.6 (2003), S. 1163–1175. DOI: 10.1016/S0165-1684(03)00037-9.

[164] T. Söderström. „Errors-in-variables methods in system identification“. In: *Automatica* 43.6 (2007), S. 939–958. DOI: 10.1016/j.automatica.2006.11.025.

[165] T. Söderström. „A generalized instrumental variable estimation method for errors-in-variables identification problems“. In: *Automatica* 47 (2011), S. 1656–1666. DOI: 10.1016/j.automatica.2011.05.010.

[166] T. Söderström. „System identification for the errors-in-variables problem“. In: *Transactions of the Institute of Measurement and Control* 34.7 (2012), S. 780–792. DOI: 10.1177/0142331211414616.

[167] T. Söderström und P. Stoica. „Comparison of some instrumental variable methods – Consistency and accuracy aspects“. In: *Automatica* 17.1 (1981), S. 101–115. DOI: 10.1016/0005-1098(81)90087-X.

[168] T. Söderström und P. Stoica. *System Identification.* Hrsg. von M.J. Grimble. Prentice Hall International, 2001.

[169] S. Solmaz, M. Akar, R. Shorten und J. Kalkkuhl. „Real-time multiple-model estimation of centre of gravity position in automotive vehicles“. In: *Vehicle System Dynamics* 46.9 (2008), S. 763–788. DOI: 10.1080/00423110701602670.

[170] S. Song, J.S. Lim, S. Baek und K.M. Sung. „Gauss Newton variable forgetting factor recursive least squares for time varying parameter tracking“. In: *Electronics Letters* 36.11 (Mai 2000), S. 988–990. DOI: 10.1049/el:20000727.

[171] B. Stenlund und F. Gustafsson. „Avoiding windup in recursive parameter estimation“. In: *Preprints of reglermöte 2002* (2002), S. 148–153.

[172] R. Tibshirani. „Regression Shrinkage and Selection via the Lasso“. In: *Journal of the Royal Statistical Society. Series B (Methodological)* 58.1 (1996), S. 267–288.

[173] A.N. Tikhonov. „Solution of incorrectly formulated problems and the regularization method“. In: *Soviet Mathematics Doklady.* Bd. 4. 1963, S. 1035–1038.

[174] W. D. Timmons, H. J. Chizeck, F. Casas, V. Chankong und P. G. Katona. „Parameter-Constrained Adaptive Control“. In: *Industrial & Engineering Chemistry Research* 36.11 (1997), S. 4894–4905. DOI: 10.1021/ie9606597.

[175] A. Trabelsi. „Automotive Reibwertprognose zwischen Reifen und Fahrbahn“. Diss. Düsseldorf: Universität Hannover, 2005.

[176] H. Tsunashima und H. Mori. „Condition monitoring of railway vehicle suspension using adaptive multiple model approach“. In: *Control Automation and Systems (ICCAS), 2010 International Conference on.* Okt. 2010, S. 584–589.

[177] H. Tsunashima, M. Murakami und J. Miyataa. „Vehicle and road state estimation using interacting multiple model approach“. In: *Vehicle System Dynamics* 44 (2006), S. 750–758. DOI: 10.1080/00423110600885772.

[178] T.Z. Um, J.G. Lee, S.T. Park und C.G. Park. „Noise covariances estimation for systems with bias states“. In: *Aerospace and Electronic Systems, IEEE Transactions on* 36.1 (Jan. 2000), S. 226–233. DOI: 10.1109/7.826324.

[179] H.J. Unrau. „Der Einfluss der Fahrbahnoberflächenkrümmung auf den Rollwiderstand, die Cornering Stiffness und die Aligning Stiffness von Pkw-Reifen“. Diss. Karlsruher Institut für Technologie, 2013. DOI: 10.5445/KSP/1000032808.

[180] A. Vahidi, A. Stefanopoulou und H. Peng. „Experiments for online estimation of heavy vehicle's mass and time-varying road grade“. In: *Proceedings of the 2003 ASME Dynamic Systems and Control Division, International Mechanical Engineering Congress and Exposition.* 2003, S. 451–458.

[181] A. Vahidi, A. Stefanopoulou und H. Peng. „Recursive least squares with forgetting for online estimation of vehicle mass and road grade: theory and experiments“. In: *Vehicle System Dynamics* 43.1 (2005), S. 31–55. DOI: 10.1080/00423110412331290446.

[182] S. Van Huffel, Hrsg. *Total least squares and errors in variables modeling : analysis, algorithms and applications.* Dordrecht: Kluwer Academic, Aug. 2002.

[183] S. Van Huffel und J. Vandewalle. „Algebraic connections between the least squares and total least squares problems“. In: *Numerische Mathematik* 55.4 (1989), S. 431–449. DOI: 10.1007/BF01396047.

[184] S. Van Huffel und J. Vandewalle. „Analysis and Properties of the Generalized Total Least Squares Problem $AX \approx B$ When Some or All Columns in A are Subject to Error“. In: *SIAM Journal on Matrix Analysis and Applications* 10.3 (1989), S. 294–315. DOI: 10.1137/0610023.

[185] S. Van Huffel und J. Vandewalle. *The total least squares problem : computational aspects and analysis.* Frontiers in applied mathematics ; 9. Philadelphia: SIAM, 1991.

[186] T. Van Waterschoot, G. Rombouts und M. Moonen. „Optimally regularized adaptive filtering algorithms for room acoustic signal enhancement“. In: *Signal Processing* 88.3 (2008), S. 594–611. DOI: 10.1016/j.sigpro.2007.09.001.

[187] P. Vandewalle, J. Kovacevic und M. Vetterli. „Reproducible research in signal processing“. In: *Signal Processing Magazine, IEEE* 26.3 (2009), S. 37–47. DOI: 10.1109/MSP.2009.932122.

[188] H.-G. Wahl und F. Gauterin. „An iterative dynamic programming approach for the global optimal control of hybrid electric vehicles under real-time constraints“. In: *Intelligent Vehicles Symposium (IV), 2013 IEEE.* Juni 2013, S. 592–597. DOI: 10.1109/IVS.2013.6629531.

[189] H.-G. Wahl, K.-L. Bauer, F. Gauterin und M. Holzapfel. „A real-time capable enhanced dynamic programming approach for predictive optimal cruise control in hybrid electric vehicles“. In: *Intelligent Transportation Systems - (ITSC), 2013 16th International IEEE Conference on.* Okt. 2013, S. 1662–1667. DOI: 10.1109/ITSC.2013.6728468.

[190] C. Wang, Z. Hu und K. Uchimura. „Precise curvature estimation by cooperating with digital road map“. In: *Intelligent Vehicles Symposium, 2008 IEEE.* Juni 2008, S. 859–864. DOI: 10.1109/IVS.2008.4621300.

[191] B.P. Welford. „Note on a Method for Calculating Corrected Sums of Squares and Products“. In: *Technometrics* 4.3 (1962), S. 419–420.

[192] L. Wilson. *The academic man: A study in the sociology of a profession.* Transaction Publishers, 1942.

[193] V. Winstead und I.V. Kolmanovsky. „Estimation of road grade and vehicle mass via model predictive control“. In: *Proc. IEEE Conf. Control Applications CCA 2005.* 2005, S. 1588–1593. DOI: 10.1109/CCA.2005.1507359.

[194] Y. Yang und W. Gao. „An Optimal Adaptive Kalman Filter“. In: *Journal of Geodesy* 80.4 (2006), S. 177–183. DOI: 10.1007/s00190-006-0041-0.

[195] V.J. Yohai. „High Breakdown-Point and High Efficiency Robust Estimates for Regression“. In: *The Annals of Statistics* 15.2 (1987), S. 642–656.

[196] Z. Yu, Y. Feng, L. Xiong und X. Wu. „Vehicle Mass Estimation for Four In-Wheel-Motor Drive Vehicle“. In: *Electrical Engineering and Control.* Hrsg. von Min Zhu. Bd. 98. Lecture Notes in Electrical Engineering. Springer Berlin Heidelberg, 2011, S. 117–125. DOI: 10.1007/978-3-642-21765-4_15.

[197] J. Ziegler, P. Bender, M. Schreiber, H. Lategahn, T. Strauss, C. Stiller, Thao Dang, U. Franke, N. Appenrodt, C.G. Keller, E. Kaus, R.G. Herrtwich, C. Rabe, D. Pfeiffer, F. Lindner, F. Stein, F. Erbs, M. Enzweiler, C. Knöppel, J. Hipp, M. Haueis, M. Trepte, C. Brenk, A. Tamke, M. Ghanaat, M. Braun, A. Joos, H. Fritz, H. Mock, M. Hein und E. Zeeb. „Making Bertha Drive; An Autonomous Journey on a Historic Route". In: *Intelligent Transportation Systems Magazine, IEEE* 6.2 (2014), S. 8–20. DOI: 10.1109/MITS.2014.2306552.

[198] E.V. Zima. „Fast parallel computation of the polynomial shift". In: *Parallel Processing Symposium, 1997. Proceedings., 11th International.* Apr. 1997, S. 402–406. DOI: 10.1109/IPPS.1997.580933.

[199] Y. Zou, S.C. Chan und T.S. Ng. „A recursive least M-estimate (RLM) adaptive filter for robust filtering in impulse noise". In: *Signal Processing Letters, IEEE* 7.11 (Nov. 2000), S. 324–326. DOI: 10.1109/97.873571.

[200] A.M. Zoubir, V. Koivunen, Y. Chakhchoukh und M. Muma. „Robust Estimation in Signal Processing: A Tutorial-Style Treatment of Fundamental Concepts". In: *Signal Processing Magazine, IEEE* 29.4 (Juli 2012), S. 61–80. DOI: 10.1109/MSP.2012.2183773.

Index

Scholarships and supervised thesis

Scholarships

I am grateful for the support granted by the House of Young Scientists allowing me to conduct parts of my research abroad. The following list provides information about three scholarships which I received from the Karlsruhe House of Young Scientists during my graduate studies.

1. Summer school scholarship, Neural Networks in Porto, 07/02/2012-07/06/2012, 850 €
2. Networking grant, department ELEC in Vrije Universiteit Brussel, 10/06/ 2013-10/25/2013, 1500 €
3. Foreign exchange scholarship, vehicle dynamics lab (VDL) in University of California Berkeley, 07/01/2014-12/22/2014, 9925 €

Supervised student thesis

[201] Y. Dursun. „Vergleich von RLS und RIV Schätzverfahren zur Identifikation von Fahrwiderstandsparametern". Student research project. Karlsruher Institut für Technologie, 2013.

[202] Z. Fuszenecker. „Robuste Behandlung von Ausreißern bei der Schätzung von Fahrzeugparametern". Master's thesis. Karlsruher Institut für Technologie, 2012.

[203] H. Hamann. „Tire Force Estimation for a Passenger Vehicle with the Unscented Kalman Filter". Master's thesis. University of California, Berkeley und Karlsruhe Institute of Technology, 2013.

[204] M. Knobloch. „Vergleich von Least-Squares Schätzverfahren zur Identifikation von Fahrwiderstandsparametern". Bachelor's thesis. Karlsruher Institut für Technologie, 2012.

[205] S. Meister. „Entwicklung einer wirtschaftlichen Methode zur Messung des Rollwiderstandes während der Fahrt“. Bachelor's thesis. Karlsruher Institut für Technologie, 2012.

[206] F. Mutter. „Abbildung von Komponenten-Kennfeldern mit neuronalen Netzen“. Master's thesis. Karlsruher Institut für Technologie, 2013.

[207] D. Peter. „Modellierung des Fahrwiderstandes“. Student research project. Karlsruher Institut für Technologie, 2012.

[208] M. Wolf. „Implementierung einer Methode zur Klassifizierung des Fahrstils bei Testfahrten“. Bachelor's thesis. Karlsruher Institut für Technologie, 2013.

Publications

Here is a list of own or coauthored publications during my time at the Karlsruhe Institute of Technology. A complete list of my publications can be found on researchgate (http://www.researchgate.net/profile/Stephan_Rhode/) and google scholar (http://scholar.google.de/citations?user=KZDir20AAAAJ).

Peer reviewed journal papers

[136] S. Rhode, K. Usevich, I. Markovsky und F. Gauterin. „A Recursive Restricted Total Least-squares Algorithm". In: *Signal Processing, IEEE Transactions on* 62.21 (Nov. 2014), S. 5652–5662. DOI: 10.1109/TSP.2014.2350959.

[209] S. Rhode, S. Hong, J.K. Hedrick und F. Gauterin. „Vehicle tractive force prediction with robust and windup-stable Kalman filters". In: *Control Engineering Practice* 46 (2016), S. 37–50. DOI: 10.1016/j.conengprac.2015.10.002.

Peer reviewed conference papers

[210] H. Hamann, J.K. Hedrick, S. Rhode und F. Gauterin. „Tire Force Estimation for a Passenger Vehicle with the Unscented Kalman Filter". In: *Intelligent Vehicles Symposium Proceedings, 2014 IEEE.* Juni 2014, S. 814–819. DOI: 10.1109/IVS.2014.6856391.

[133] S. Rhode, F. Bleimund und F. Gauterin. „Recursive Generalized Total Least Squares with Noise Covariance Estimation". In: *19th IFAC World Congress.* Bd. 19. International Federation of Automatic Control, 2014, S. 4637–4643. DOI: 10.3182/20140824-6-ZA-1003.01568.

[134] S. Rhode und F. Gauterin. „Vehicle mass estimation using a total least-squares approach". In: *Intelligent Transportation Systems (ITSC), 2012 15th International IEEE Conference on.* Sep. 2012, S. 1584–1589. DOI: 10.1109/ITSC.2012.6338638.

[135] S. Rhode und F. Gauterin. „Online estimation of vehicle driving resistance parameters with recursive least squares and recursive total least squares“. In: *Intelligent Vehicles Symposium (IV), 2013 IEEE.* Juni 2013, S. 269–276. DOI: 10.1109/IVS.2013.6629481.

Patents

[211] F. Bleimund und S. Rhode. „Verfahren, Computerprogrammprodukt, Vorrichtung und Fahrzeug zur Berechnung einer Variable zum Betreiben eines Fahrzeugs“. Pat. 13006043. 2013. URL: https://depatisnet.dpma.de/DepatisNet/depatisnet?action=bibdat&docid=EP000002886409A1.

Karlsruher Schriftenreihe Fahrzeugsystemtechnik (ISSN 1869-6058)

Herausgeber: FAST Institut für Fahrzeugsystemtechnik

Band 1 Urs Wiesel
Hybrides Lenksystem zur Kraftstoffeinsparung im schweren Nutzfahrzeug. 2010
ISBN 978-3-86644-456-0

Band 2 Andreas Huber
Ermittlung von prozessabhängigen Lastkollektiven eines hydrostatischen Fahrantriebsstrangs am Beispiel eines Teleskopladers. 2010
ISBN 978-3-86644-564-2

Band 3 Maurice Bliesener
Optimierung der Betriebsführung mobiler Arbeitsmaschinen. Ansatz für ein Gesamtmaschinenmanagement. 2010
ISBN 978-3-86644-536-9

Band 4 Manuel Boog
Steigerung der Verfügbarkeit mobiler Arbeitsmaschinen durch Betriebslasterfassung und Fehleridentifikation an hydrostatischen Verdrängereinheiten. 2011
ISBN 978-3-86644-600-7

Band 5 Christian Kraft
Gezielte Variation und Analyse des Fahrverhaltens von Kraftfahrzeugen mittels elektrischer Linearaktuatoren im Fahrwerksbereich. 2011
ISBN 978-3-86644-607-6

Band 6 Lars Völker
Untersuchung des Kommunikationsintervalls bei der gekoppelten Simulation. 2011
ISBN 978-3-86644-611-3

Band 7 3. Fachtagung
Hybridantriebe für mobile Arbeitsmaschinen. 17. Februar 2011, Karlsruhe. 2011
ISBN 978-3-86644-599-4

Die Bände sind unter www.ksp.kit.edu als PDF frei verfügbar oder als Druckausgabe bestellbar.

Karlsruher Schriftenreihe Fahrzeugsystemtechnik (ISSN 1869-6058)

Herausgeber: FAST Institut für Fahrzeugsystemtechnik

Band 8 Vladimir Iliev
Systemansatz zur anregungsunabhängigen Charakterisierung des Schwingungskomforts eines Fahrzeugs. 2011
ISBN 978-3-86644-681-6

Band 9 Lars Lewandowitz
Markenspezifische Auswahl, Parametrierung und Gestaltung der Produktgruppe Fahrerassistenzsysteme. Ein methodisches Rahmenwerk. 2011
ISBN 978-3-86644-701-1

Band 10 Phillip Thiebes
Hybridantriebe für mobile Arbeitsmaschinen. Grundlegende Erkenntnisse und Zusammenhänge, Vorstellung einer Methodik zur Unterstützung des Entwicklungsprozesses und deren Validierung am Beispiel einer Forstmaschine. 2012
ISBN 978-3-86644-808-7

Band 11 Martin Gießler
Mechanismen der Kraftübertragung des Reifens auf Schnee und Eis. 2012
ISBN 978-3-86644-806-3

Band 12 Daniel Pies
Reifenungleichförmigkeitserregter Schwingungskomfort – Quantifizierung und Bewertung komfortrelevanter Fahrzeugschwingungen. 2012
ISBN 978-3-86644-825-4

Band 13 Daniel Weber
Untersuchung des Potenzials einer Brems-Ausweich-Assistenz. 2012
ISBN 978-3-86644-864-3

Band 14 **7. Kolloquium Mobilhydraulik. 27./28. September 2012 in Karlsruhe.** 2012
ISBN 978-3-86644-881-0

Band 15 4. Fachtagung
Hybridantriebe für mobile Arbeitsmaschinen 20. Februar 2013, Karlsruhe. 2013
ISBN 978-3-86644-970-1

Die Bände sind unter www.ksp.kit.edu als PDF frei verfügbar oder als Druckausgabe bestellbar.

Karlsruher Schriftenreihe Fahrzeugsystemtechnik (ISSN 1869-6058)

Herausgeber: FAST Institut für Fahrzeugsystemtechnik

Band 16 Hans-Joachim Unrau
Der Einfluss der Fahrbahnoberflächenkrümmung auf den Rollwiderstand, die Cornering Stiffness und die Aligning Stiffness von Pkw-Reifen. 2013
ISBN 978-3-86644-983-1

Band 17 Xi Zhang
Untersuchung und Entwicklung verschiedener Spurführungsansätze für Offroad-Fahrzeuge mit Deichselverbindung. Noch nicht erschienen
ISBN 978-3-7315-0005-6

Band 18 Stefanie Grollius
Analyse des gekoppelten Systems Reifen-Hohlraum-Rad-Radführung im Rollzustand und Entwicklung eines Rollgeräuschmodells. 2013
ISBN 978-3-7315-0029-2

Band 19 Tobias Radke
Energieoptimale Längsführung von Kraftfahrzeugen durch Einsatz vorausschauender Fahrstrategien. 2013
ISBN 978-3-7315-0069-8

Band 20 David Gutjahr
Objektive Bewertung querdynamischer Reifeneigenschaften im Gesamtfahrzeugversuch. 2014
ISBN 978-3-7315-0153-4

Band 21 Neli Ovcharova
Methodik zur Nutzenanalyse und Optimierung sicherheitsrelevanter Fahrerassistenzsysteme. 2014
ISBN 978-3-7315-0176-3

Band 22 Marcus Geimer, Christian Pohlandt
Grundlagen mobiler Arbeitsmaschinen. 2014
ISBN 978-3-7315-0188-6

Band 23 Timo Kautzmann
Die mobile Arbeitsmaschine als komplexes System. 2014
ISBN 978-3-7315-0187-9

Die Bände sind unter www.ksp.kit.edu als PDF frei verfügbar oder als Druckausgabe bestellbar.

Karlsruher Schriftenreihe Fahrzeugsystemtechnik (ISSN 1869-6058)

Herausgeber: FAST Institut für Fahrzeugsystemtechnik

Band 24 Roman Weidemann
Analyse der mechanischen Randbedingungen zur Adaption der oszillierenden Hinterschneidtechnik an einen Mobilbagger. 2014
ISBN 978-3-7315-0193-0

Band 25 Yunfan Wei
Spurführungsregelung eines aktiv gelenkten Radpaars für Straßenbahnen. 2014
ISBN 978-3-7315-0232-6

Band 26 David Schmitz
Entwurf eines fehlertoleranten Lenkventils für Steer-by-Wire Anwendungen bei Traktoren. 2014
ISBN 978-3-7315-0264-7

Band 27 Christian Schwab
Beitrag zu einer universellen Baggerschnittstelle zur Übertragung elektrischer und hydraulischer Leistung sowie elektronischer Signale für komplexe Anbaugeräte. 2014
ISBN 978-3-7315-0281-4

Band 28 Peter Dengler
Untersuchung zum effizienten Betrieb von Hydraulikzylindern in Konstantdrucksystemen unter Verwendung einer Zwischendruckleitung. 2015
ISBN 978-3-7315-0295-1

Band 29 Manuel Bös
Untersuchung und Optimierung der Fahrkomfort- und Fahrdynamikeigenschaften von Radladern unter Berücksichtigung der prozessspezifischen Randbedingungen. 2015
ISBN 978-3-7315-0310-1

Band 30 5. Fachtagung
Hybride und energieeffiziente Antriebe für mobile Arbeitsmaschinen
25. Februar 2015, Karlsruhe. 2015
ISBN 978-3-7315-0323-1

Die Bände sind unter www.ksp.kit.edu als PDF frei verfügbar oder als Druckausgabe bestellbar.

Karlsruher Schriftenreihe Fahrzeugsystemtechnik (ISSN 1869-6058)

Herausgeber: FAST Institut für Fahrzeugsystemtechnik

Band 31 Michael Eckert
Energieoptimale Fahrdynamikregelung mehrmotoriger Elektrofahrzeuge. 2015
ISBN 978-3-7315-0332-3

Band 32 Martin Scherer
Beitrag zur Effizienzsteigerung mobiler Arbeitsmaschinen. Entwicklung einer elektrohydraulischen Bedarfsstromsteuerung mit aufgeprägtem Volumenstrom. 2015
ISBN 978-3-7315-0339-2

Band 33 Rinaldo Arnold
Automatische Abstimmung der Sekundärseite eines dreiphasigen Systems zur berührungslosen induktiven Energieübertragung. 2015
ISBN 978-3-7315-0355-2

Band 34 Johannes Gültlinger
Kraftübertragung und Fahrbahnverschleiß durch Spikereifen. 2015
ISBN 978-3-7315-0358-3

Band 35 Thorsten Dreher
Energieeffizienz von Konstantdrucksystemen mit sekundärgeregelten Antrieben beim Einsatz in mobilen Arbeitsmaschinen. 2015
ISBN 978-3-7315-0377-4

Band 36 Steffen Kölling
Konzeptionelle Untersuchung zur Neigekompensation von Stromabnehmern. 2015
ISBN 978-3-7315-0387-3

Band 37 Michael Fritz
Entwicklungswerkzeuge für die Fahrzeugklimatisierung von Nutzfahrzeugen. 2015
ISBN 978-3-7315-0384-2

Karlsruher Schriftenreihe Fahrzeugsystemtechnik (ISSN 1869-6058)

Herausgeber: FAST Institut für Fahrzeugsystemtechnik

Band 38 Ralf Oberfell
Stochastische Simulation von Energieflüssen im Nutzfahrzeug. Ein einsatzorientiertes Bewertungs- und Optimierungsverfahren. 2015
ISBN 978-3-7315-0403-0

Band 39 Christoph Sturm
Bewertung der Energieeffizienz von Antriebssystemen mobiler Arbeitsmaschinen am Beispiel Bagger. 2015
ISBN 978-3-7315-0404-7

Band 40 Florian Netter
Komplexitätsadaption integrierter Gesamtfahrzeugsimulationen. 2016
ISBN 978-3-7315-0414-6

Band 41 Markus Springmann
Auslegung eines asynchronen Langstatorlinearmotors mit großem Luftspalt als Straßenbahnantrieb. 2015
ISBN 978-3-7315-0418-4

Band 42 Alexander Basler
Eine modulare Funktionsarchitektur zur Umsetzung einer gesamtheitlichen Betriebsstrategie für Elektrofahrzeuge. 2015
ISBN 978-3-7315-0421-4

Band 43 Hans-Georg Wahl
Optimale Regelung eines prädiktiven Energiemanagements von Hybridfahrzeugen. 2015
ISBN 978-3-7315-0422-1

Band 44 Jennifer Heck
Zur Simulation des Rad-Schiene-Verschleißes bei Straßenbahnen. 2016
ISBN 978-3-7315-0443-6

Die Bände sind unter www.ksp.kit.edu als PDF frei verfügbar oder als Druckausgabe bestellbar.

Karlsruher Schriftenreihe Fahrzeugsystemtechnik (ISSN 1869-6058)

Herausgeber: FAST Institut für Fahrzeugsystemtechnik

Band 45	Moritz Vaillant **Design Space Exploration zur multikriteriellen Optimierung elektrischer Sportwagenantriebsstränge: Variation von Topologie und Komponenteneigenschaften zur Steigerung von Fahrleistungen und Tank-to-Wheel Wirkungsgrad.** 2016 ISBN 978-3-7315-0452-8
Band 46	Philip Nagel **Entwicklung einer Betriebsstrategie zur Energierückgewinnung in hybriden Mehrverbrauchersystemen.** 2016 ISBN 978-3-7315-0479-5
Band 47	Matthias Pfriem **Analyse der Realnutzung von Elektrofahrzeugen in kommerziellen Flotten zur Definition einer bedarfsgerechten Fahrzeugauslegung.** 2016 ISBN 978-3-7315-0489-4
Band 48	Mohanad El-Haji **Ontologie-basierte Definition von Anforderungen an Validierungswerkzeuge in der Fahrzeugtechnik.** 2016 ISBN 978-3-7315-0496-2
Band 49	**9. Kolloquium Mobilhydraulik 22./23. September 2016 in Karlsruhe.** 2016 ISBN 978-3-7315-0573-0
Band 50	6. Fachtagung **Hybride und energieeffiziente Antriebe für mobile Arbeitsmaschinen 15. Februar 2017, Karlsruhe.** 2017 ISBN 978-3-7315-0601-0
Band 51	Fabian Schirmaier **Experimentelle Untersuchung und Simulation des Umformverhaltens nähgewirkter unidirektionaler Kohlenstofffasergelege.** 2017 ISBN 978-3-7315-0620-1

Die Bände sind unter www.ksp.kit.edu als PDF frei verfügbar oder als Druckausgabe bestellbar.

Karlsruher Schriftenreihe Fahrzeugsystemtechnik (ISSN 1869-6058)

Herausgeber: FAST Institut für Fahrzeugsystemtechnik

Band 52 Mathias Cabrera Cano
Neuronale Netze mit externen Laguerre-Filtern zur automatischen numerischen Vereinfachung von Getriebemodellen. 2017
ISBN 978-3-7315-0621-8

Band 53 Arwed Schmidt
Flottenbetrieb von elektrischen und autonomen Serviceagenten im städtischen Personennahverkehr. 2017
ISBN 978-3-7315-0633-1

Band 54 Katharina Knaisch
Untersuchung von Spulensystemen zur induktiven Energieübertragung von Elektrofahrzeugen. Vergleich von Topologien und Entwicklung einer Auslegungsmethodik. 2017
ISBN 978-3-7315-0630-0

Band 55 Frank Christof Stalter
Ansätze zur akustischen Optimierung von Reifen und Fahrbahnen für Elektrofahrzeuge unter Antriebsmoment. 2017
ISBN 978-3-7315-0645-4

Band 56 Steffen Rose
Modellbildung und Simulation von mobilen Arbeitsmaschinen. Untersuchungen zu systematischen Modellvereinfachungen in der Simulation von Antriebssystemen am Beispiel Bagger. 2017
ISBN 978-3-7315-0684-3

Band 57 Ulrico Peckelsen
Objective Tyre Development. Definition and Analysis of Tyre Characteristics and Quantification of their Conflicts. 2017
ISBN 978-3-7315-0713-0

Band 58 Stefan Haag
Sequentieller Versuch zur HiL-unterstützten Validierung hybrider Antriebssysteme mit gekoppelten Antriebseinheiten. 2018
ISBN 978-3-7315-0752-9

Karlsruher Schriftenreihe Fahrzeugsystemtechnik (ISSN 1869-6058)

Herausgeber: FAST Institut für Fahrzeugsystemtechnik

Band 59 Dirk Hülsebusch
Fahrerassistenzsysteme zur energieeffizienten Längsregelung - Analyse und Optimierung der Fahrsicherheit. 2018
ISBN 978-3-7315-0755-0

Band 60 Christian Pohlandt
Intelligentes Gesamtmaschinenmanagement für elektrische Antriebssysteme. 2018
ISBN 978-3-7315-0774-1

Band 61 Oliver Maier
Modellbasierte Entwicklung eines aktiven Sicherheitssystems für elektrifizierte Fahrräder. 2018
ISBN 978-3-7315-0778-9

Band 62 Stephan Rhode
Robust and Regularized Algorithms for Vehicle Tractive Force Prediction and Mass Estimation. 2018
ISBN 978-3-7315-0807-6

Die Bände sind unter www.ksp.kit.edu als PDF frei verfügbar oder als Druckausgabe bestellbar.